Marketing Projects

Best Practices in Portfolio, Program, and Project Management

Ginger Levin

RECENTLY PUBLISHED TITLES

Implementing Project and Program Benefit Management
Kenn Dolan

Culturally Tuning Change Management
Risto Gladden

The Four Pillars of Portfolio Management: Organizational Agility, Strategy, Risk, and Resources
Olivier Lazar

Systems Engineering for Projects: Achieving Positive Outcomes in a Complex World
Lory Mitchell Wingate

The Human Factor in Project Management
Denise Thompson

Project Business Management
Oliver F. Lehmann

PgMP® Exam Test Preparation: Test Questions, Practice Tests, and Simulated Exams
Ginger Levin

Managing Complex Construction Projects: A Systems Approach
John K. Briesemeister

Managing Project Competence: The Lemon and the Loop
Rolf Medina

The Human Change Management Body of Knowledge (HCMBOK®), Third Edition
Vicente Goncalves and Carla Campos

Creating a Greater Whole: A Project Manager's Guide to Becoming a Leader
Susan G. Schwartz

Marketing Projects

Olivier Mesly

CRC Press
Taylor & Francis Group
Boca Raton London New York

CRC Press is an imprint of the
Taylor & Francis Group, an **informa** business

AN AUERBACH BOOK

CRC Press
Taylor & Francis Group
6000 Broken Sound Parkway NW, Suite 300
Boca Raton, FL 33487-2742

First issued in paperback 2022

© 2020 by Taylor & Francis Group, LLC
CRC Press is an imprint of Taylor & Francis Group, an Informa business

No claim to original U.S. Government works

ISBN 13: 978-1-03-247480-9 (pbk)
ISBN 13: 978-1-138-19787-9 (hbk)

DOI:10.1201/9781315272252

To my beloved daughter, Alexandra.

Congratulations on your academic achievements.

Contents

List of Abbreviations and Acronyms

4Ps of marketing	Product, Price, Promotion, Place
4Ps of project management	Plan, Processes, People, Power
6Ps	The six angles of analysis of strategic project management: 4Ps, PRO, POVs, POE, POW, PWP
AICIEP	Awareness, Interest, Consideration, Intent, Evaluation, and Purchasing
AIDA	Attention, Interest, Desire, Action
B2B	Business-to-Business exchange
B2C	Business-to-Consumer exchange
BOK	Book of knowledge
CLV	Customer Lifetime Value
CPWB	Counterproductive work behaviors
CRM	Customer Relationship Management
CSR	Corporate social responsibility
DC	Decision to buy and consume
DS	Dominant strategy
DSM-IV (or DSM-V)	*Diagnosis and Statistical Manual of Mental Disorders*
EVM	Earned Value Management
FAST	Functional analysis system technique
FMEA	Failure Mode and Effect Analysis
FTA	Fault Tree Analysis
GANTT	A charting system named after American engineer, Henry Gantt
JCCBI	Jacques Cartier and Champlain Bridges Incorporated
KCFs	Key Consensus Factors or Key Consensus Fundamentals

KFFs	Key Failure Factors or Key Failure Fundamentals
KSFs	Key Success Factors or Key Success Fundamentals
MBTI	Myers-Briggs personality test
MFP	Marketing Feasibility of Projects
MSDS	Material Safety Data Sheet
OBS	Organization Breakdown Structure
OPA	Optimal Path Analysis
PERT	Program Evaluation and Review Technique
PBS	Product Breakdown Structure
PMBOK	Project Management Body of Knowledge
PMI	Project Management Institute
PMP	Project management professional
POC	Proof of Concept
POE	Point of equilibrium
POV	Point of vulnerability
POW[1]	Product, Organization, Work
POW	Prisoners of war
PPP	Private-Public Partnerships
PRO	Pessimistic, Realistic, Optimistic scenarios
PWP	Work psychodynamics
R&D	Research and Development
SKUs	Stock keeping units
SVOR	Strengths, vulnerabilities, opportunity, risks
SWOT	Strengths, weaknesses, opportunity, threats
WBS	Work Breakdown Structure
WIP	Work In Progress

---

[1] This is in not related to "Prisoner of War" (POW).

Preface

When Walt Disney conceived Mickey Mouse, he had a project in mind: He wanted to create something memorable that would appeal to children. He had yet to market this project, but that was to come. Recognized the world over, Mickey Mouse's ears speak to what Disney achieved. Most mice have small ears proportionate to their bodies, not ones that resemble the ears of an elephant. Yet, Disney designed Mickey with big, round ears that emulate satellite dishes. In other words, Mickey Mouse listens. Besides candy or occasionally "taking the mickey" (making fun) out of each other, there is hardly anything children like more than being listened to. Disney hit the jackpot. Let us be like him; let us conceive products; and, more particularly, let us conceive projects that materialize into deliverables (products or services in particular) that meet end customers' enduring needs!

It is also important to be aware of what a project is fundamentally about. The front cover is an artistic rendering of a set of bicycle cogs, the picture for which is on the back cover. At first glance, one cannot see the intricate mechanisms that compose the painting, but after examining the image on the back and understanding the underlying structures, one can see how the artist fools the eyes, so to speak. Projects are the same; behind every aspect, there lies an underlying truth.

Notes on the Book

Thank you for acquiring this book on the *marketing feasibility of projects*. We hope it facilitates your understanding of both marketing and project management and assists you in making your innovative products a success. Below are a few notes to keep in mind while using this text:

1. we have used gender-neutral language to adopt a neutral position;
2. we have used **bold face** to highlight a term or new concept, or to isolate it from the text to ease comprehension;
3. we have made our translations from French to English identifiable;
4. we have used American English;

5. we have used an alphabetical order for organizing listings, except where it would compromise the meaning of the sentence;
6. we have defined acronyms as often as needed, in order to ease reading and facilitate comprehension;
7. unless indicated otherwise, some acronyms are set in the plural form when it eases reading. (for example, Key Success Factors is KSFs);
8. we have *italicized* foreign words;
9. names of people and places are real (unless indicated otherwise);
10. we have identified constructs with a capital letter; if the same word is later lowercased it is meant in its normal context (for example, when we capitalize "trust"—Trust—we refer to a construct);
11. we have repeated some tables, graphs, and expressions on a limited number of occasions to facilitate ease of reading and eliminate the need to flip back and forth through the text;
12. we have drawn curves in tables and figures in a simplified, stylized manner for ease of reading;
13. we can make the data set and other questionnaires on psychological constructs available upon request;
14. we do not mean to be derogatory when making remarks or critiques; rather, they should be taken constructively and as a commendable effort to increase knowledge;
15. we have made the best effort to substantiate hypotheses and express findings in the most accurate manner;
16. we have corrected the English on some of the texts we have quoted to make the sentences' meaning clearer.

Note on the Websites

This book refers to over 20 exercises aimed at helping readers develop the analytical skills necessary to commence a marketing feasibility analysis of projects. All are straightforward, and are discussing during our seminars.

Acknowledgments

We would like to thank many people and organizations for their kind and generous support in writing this book.

We are grateful to the management at ICN Business School, which allocated a most helpful budget to fund our endeavor. Many others at ICN also contributed, giving their time and providing technical support, including Estelle Durand and Justine Kayinamura. I also thank Silvester and Vera Ivanaj, among many other colleagues. Jennifer Dinsmore (of Jennifer Dinsmore Editorial) revised and corrected our text and did a wonderful job.

We thank the various business people who have given their time generously over the past few years and who have provided fabulous testimonials, sometimes putting concepts into words better than any book on marketing or project management has ever done. Mr. Christophe Oliveira has been a key resource in directing us toward managers willing to participate.

We are also very thankful to the students who agreed to provide some of their work, thus allowing readers a glimpse as to how theory can be put into practice. We owe a special, grateful mention to the friends who supported our work over the years, including Maria Arruda, Stéphane Bouchard, Sophie Boyer, Alain Desrochers, Miladin Djerkovic, Jean-Marc Ladouceur, Sarah Moser, Kofi Nyamekye, and Michel Paiement.

We'd like to thank Darloz (Facebook: DarlozArt) for providing the images, including cover image.[1]

Many other participants from various countries, including Canada, France, and Luxembourg, provided testimonials and shared their thoughts and experiences, without counting the hours. They enriched our text with real-word examples so it may capture not only the theory behind the marketing feasibility of projects, but also its day-to-day applications. We feel these complement the book very nicely indeed, and they may be found in Chapter 6 as well as in the exercises supplied during our seminars.

---

[1] Notes: Except where indicated, photos and drawings are by artist Darloz.

Authors

Dr. Olivier Mesly teaches at ICN Business School and University of Lorraine, Nancy, France, with marketing and project management being his topics of predilection. He also acts as a consultant for a variety of organizations and government-sponsored offices and is the author of numerous books on project feasibility and scholarly articles in English, French, and Spanish.

As head of marketing for large foreign and Canadian-owned companies, over the years Olivier has launched new products in many markets. As a researcher, he has researched over 50 different teams assigned to a wide variety of projects and acts (and has acted) as a consultant for various organizations—from local, government-sponsored development agencies to independent businesses (e.g., Kruger). He completed his postdoctoral fellowship at HEC Montréal, graduated with a doctor of business administration in the record time of two-and-a-half years, and holds an MBA in international marketing from Guelph University. He also has a BA, with distinction, in Japanese Studies and a diploma in Public Relations from McGill University. As of April 2017, he also has a certificate in vocal music (classical singing) from the University of Sherbrooke in Canada.

Olivier has written several books, numerous marketing case studies, and exercises for feasibility studies, as well as more than 30 peer-reviewed articles published in major English, French, and Spanish journals (a brief summary of his publications can be found in the reference section.)

Michel Makiela, PhD, is an associate professor of Marketing at ICN Business School, where he heads the Executive MBA program. His focus is on innovation in business models, and before joining ICN, he acted as the director of marketing and new business development for multinational companies (including Dassault); he also has experience in the automotive industry as a consultant (e.g., PSA).

Sybille Persson, PhD, is a professor at ICN Business School in the Human Resource and Organizational Behavior Department and is a member of CEREFIGE (University of Lorraine). She is founder of the ICN School of Coaching and, for the past 30 years, has been heavily involved in guiding firm managers through training, consulting, coaching, and team building. She has published numerous academic and professional articles on human resource management and strategy.

François Trochu, PhD, is professor of Mechanical Engineering at Polytechnique Montréal. He has developed original software for composite manufacturing by resin injection, which is used commercially worldwide. Former holder of a Tier I Canada Research Chair on high performance composites, he obtained two Industrial Research Chairs with General Motors and Safran. He has published 130 papers in scientific journals and presented more than 100 refereed lectures at international conferences. His current focus is on success in innovation.

Chapter I
General Introduction

Located in Dublin, Ireland, this building was specifically designed to contrast with the "normal" way of doing things. That's one way of marketing the town.

Nothing ever is wasted, but everything is forever reinvented.

Mesly (2019)

Determining the marketing feasibility of projects (MFP) is about asking six questions:

1. Does the project aim at putting an innovative product in the market?
2. Will the product answer the needs of the intended market segments?
3. Have the managers measured the product (the deliverable) against its competitors, for its lifecycle, and against its family of products?
4. Have the managers put forth a sound strategy with respect to the product itself, its price, its promotion, and its intended distribution?
5. In what way will the consumers/end users perceive the value, and different attributes, of the product?
6. How will repeated consumption (or use) of the product be guaranteed?

Convincing answers will lead a marketing feasibility expert to sway in favor of the project, based on a proper marketing analysis. Technical and financial (e.g., forecasting sales) analyses are usually necessary, and all of them interact to lead the expert to a decision. Analysts will make one of three:

1. Go with the project →
2. Conditional go ←
3. No-go ↓

A feasibility analysis, regardless of marketing concerns and, in simple terms, sets out to determine whether marketers have the means and forces to materialize the idea.

A marketing feasibility analysis is not only about what marketing experts are going to do once managers materialize their projects, at which time the latter become operations; nor is it about forecasting sales, as this relates to products once they are in the marketplace and belongs to a financial analysis. Indeed, feasibility studies include analyses of expected sales, but these are financial analyses and must be supported by careful estimations not by wild hopes about a product before it has been created (e.g., once the project is completed and is truly an operation) but by sound marketing research. Marketing projects is at the junction of marketing and project management knowledge.

Business has changed and now requires such an approach. When we (the authors of this book) first discuss business with some colleagues or students, we find few who have a lot to say about the marketing projects, but this is through no fault of theirs. In fact, based on our combined teaching experiences in Canada, Latin America, Asia, and Europe, courses in marketing feasibility seldom discuss project management, and many replicate theories that pertain to the marketing field in an attempt to make them fit into the title of the course—without any deeper thought. However, that does not do justice to how the business world evolves, or to how managers should conduct marketing feasibility analyses.

Managers must take a number of complex operations into consideration, including ecological (e.g., carbon footprint) and ergonomic ones, as well as sustainable development, work conditions, efficiency and effectiveness in production, economic and financial returns, and community engagement and governance. They must decide whether the product is the star of the business, or the customers and the salespeople they employ who should be the business priority. New products attract attention, but as time goes on, service becomes increasingly important, that is, the ability to maintain pleasant relationships between marketing agents. Managers must balance market growth potential and market share, which is the slice of the market their business occupies. A healthy combination of prospect for growth and a large market share (something Apple and Samsung do well[1]) is ideal. This requires measuring both the competition and customers' states of mind, heart, and wallet on a personal level, while measuring the market share on a social level.

Projects have swept over the world like hurricanes (especially those on digital platforms), and new products or fancy new infrastructures mushrooming everywhere you look. Indeed, more and more companies operate based on projects. To mention only one sector of economic activity, between 2012 and 2022, studies forecast that the United States alone will hire some 78,200 new construction managers.[2] Residential and non-residential construction in 2015 amounted to $612 billion USD, with the former representing slightly more than 50% of the total. Since the end of World War II, the World Bank has supported more than 13,000 projects in over 173 countries.[3] Projects respect social, legal, and regulatory conditions, which range from solving structural problems (such as for a bridge) to providing water in remote locations (such as in South America). In many governmental organizations and large companies, top management put strategies into place which they then divide into **project portfolios** (much like financial experts handle portfolios of their own). Sometimes, these are sub-divided into programs and combine projects of a similar nature.

In order to evolve, societies must create projects. Most often, promoters have to present them to potential investors (Note: Project promoters are not necessarily the same people as the marketers: They often are the initial visionaries of the project who then hire expert marketers to market their projects). In other words,

---

[1] For more, see the Boston Consulting Group's (BCG) analytical framework.
[2] PM Network magazine (June 15, p. 17).
[3] See www.worldbank.org/projects. Accessed August 5, 2015.

even before projects start, promoters must market them; that is, present them in an appealing way. Once project managers complete their projects, they must then present them to end users so the deliverables actually serve their initial purpose and generate revenues. Investors do not finance projects if they do not fulfill a need and, as such, projects are a pure expression of marketing—to put something in the market that marketers assume the market needs.

Therefore, we can say that marketing and project management go hand in hand. Yet there is still more to it. We have at times seen promoters presenting projects to potential investors, or seen eager students developing a marketing plan, regardless of costs and without taking in their implications. In reality, even before one goes ahead with preparing a full-fledged marketing plan, one must decide whether the project is feasible. Project feasibility encompasses many aspects, among which are technical and financial considerations. Marketers have no choice but to consider these aspects, even though their focus is actually on fulfilling the needs of the projects' promoters and/or the end users'. Marketing projects is about determining if it makes sense to go ahead with a project by using analytical tools pertaining to marketing, such as needs analyses, and adding management concepts such as the triple constraints of costs, calendars of tasks and activities, and quality parameters. It is a noble venture, one needed in our modern world of high volatility.

Certainly, this book has the advantage of preparing marketers and project managers, as well as students, for the contemporary business world, one where marketing efforts are, in most cases, articulated around projects, which cannot lift off without marketing knowledge. Nowadays, marketers are indeed project managers in their own rights, and project managers can only make sense of their role if they understand the foundations of marketing. In addition, before both engage in any efforts that could cost their companies substantial amounts of money, they must take a break and analyze their business model to determine, as scientifically as possible, if these efforts are feasible. In short, they must ask themselves: Will the company developing the project be able to market it? In addition, is the company promoting a solid project that will indeed respond to end users' needs?

All projects have a beginning and an end, after which they become normal operations. They, theoretically, have a cost ceiling and minimal (floor) quality requirements, and all offer something unique and innovative (hence the necessity to resort to various marketing tools). They potentially affect human lives one at a time, or else by groups of tens or hundreds or thousands or millions. Unlike hurricanes, however, projects actually serve humanity and improve the human condition—well, most of them anyway.

Why do we require a cost feasibility analysis? Because products are increasingly sophisticated and resort, more and more, to a variety of technologies that make them expensive to develop. Business investors want to recover their investments and certainly earn a return. Regretfully, many people commence projects (or student assignments) without proper assessment as to whether they make sense in the first place. Our reality has become so multifaceted that we throw ourselves into the first idea that appeals to us, without due consideration of a multitude of factors—a sure recipe for disaster.

The term **project capability** is one way of explaining our viewpoint. It refers to "the appropriate knowledge, experience, and skills necessary to perform pre-bid, bid, project, and post-project activities."[4] Project Management Book of Knowledge (PMBOK) divides these skills into three categories, named the Project Management Institute (PMI) Talent Triangle: leadership, technical, and managerial. We would argue that, given the foundation of this book, these skills also include marketing endeavors, such as working with clients and communicating with stakeholders.

A number of factors threaten a project's feasibility, especially from a marketing point of view. We will not pretend to list them all, but let us take a quick look at some realities project managers, investors, governments, end users, and marketers face. These realities all have two faces: one good and the other challenging to say the least. Four immediately come to mind: relentless innovation, the influence of the Internet, virtual teams (which are developing fast,[5] and with them, a load of communication challenges), and resistance.

First, as noted above, our modern world revolves around relentless innovation. Every year, thousands of new products invade the markets and patent numbers explode on a global scale. Those countries that develop the most (or those that copy innovative products, whether legally or illegally) are those that tend to dominate or fare well in the world's economy, including South Korea, Japan, Switzerland, Germany, and the United States.

Managers must turn all these ideas into projects that will materialize, and then market them well to end users. The upside is that each innovation is supposed to improve our living conditions; the downside is that the lifecycle of products is diminishing, resulting in rapid product obsolescence and over-consumerism.

Second, another reality that did not exist twenty years ago—if we even need to go that far back—is the presence and influence of the World Wide Web. A wealth of information is always available, but the downside is this increases the volatility of the market, not to mention that it is full of erroneous, self-centered data that does not serve the common good. In fact, the overload of such information has not helped humankind grow smarter, but rather it has encouraged us to become lazy, trustful of illusions, and prompt to be (willingly or unwillingly) deceitful or deceived.

Third, along with the surge in novel products and dissemination across the World Wide Web, comes the obligation to work in teams. This is especially true given the complexity of many projects and, with it, the possibility of working from home or from remote locations. Virtual teams are great, as they may foster virtual friendships, but they have the potential to create big holes in project management. Distance matters and may even generate hostility.[6] Many times, projects have failed

---

[4] Davies and Brady (2000, p. 62).
[5] Clark and Wheelwright (1992), Hobday (2000).
[6] Milgram (1974).

because of communication lapses; some engineers or project managers forgot to tell the other something of significance. The fact that team members work far apart leads them to forget about minute details, which then blossom into disasters because they are vital parts of a project. A butterfly effect of sorts has been observed.

Finally, resistance is another factor plaguing the evolution of marketing and project management. Employers expect a full commitment from employees, but too many opportunities invite them to do just the opposite. Who does not use work time to check Facebook or search for the next thing to buy on Amazon? Resistance comes in all kinds of forms. In terms of planning, one can almost count on opacity on the part of top managers, as they tend to hide or withhold critical information; in regard to processes, delaying, stalling, or procrastinating, as well as obstruction, inflame working relationships, especially in economies more prone to individualism. In terms of people, negative behaviors include denigrating others, arguing uselessly, spreading rumors, and similar behaviors. In terms of use of authority, one can imagine having difficulties with some managers not complying with regulations (corruption, tax evasion, etc.) or deliberately misguiding the company in favor of short-term profits (often through short-term benefits granted to the fearless leaders!).

Therefore, if we are to take a look at our work as it stands now, from the viewpoint of marketing projects, we face a somewhat promising yet challenging situation (Figure I.1).

Figure I.1   The contemporary challenges of marketers and project managers.

Note: Marketing and project managers face common challenges: An increasing level of innovation (and complexity) that soon turns their most recent efforts into obsolescence. This forces them to constantly reexamine the customers' motivations; the ease of access to the Internet (which, unfortunately, forms strong but ellusive opinions based on erroneous facts, making consumers more, yet wrongly, informed); the presence of virtual networks and teams, which cuts into the depth of trust and effectiveness of analyses; and, finally, resistance, rendered easy by increasing levels of means to deviate from project and marketing plans (strikes, falsifying documents, etc.) with little repercussions.

When facing these challenges, marketing and project managers also have other considerations that pertain to their own areas of expertise. Both must agree on a full and accurate definition of their common project and plan judiciously, explicitly defining the input and output and determining the transformation phases of the project in terms of the calendar of tasks and activities, costs, and norms of quality. They must then gather the necessary resources, lay out the value proposed to their customers, face potential risks, reduce vulnerabilities, and minimize management and marketing errors, adapting their efforts in both short term and long term, and weeding out unwanted elements (e.g., individuals or groups who intend to hinder the project). Most of all, they must sustain stakeholders' trust in the project.

These are not easy tasks, and only through mutual understanding can marketers (who are often salespeople in their own right, as well as statisticians) and project managers (who are often engineers or human resource-oriented professionals) proceed to materialize the product, market it, and make it a success. The road they travel together will be filled with sparks that will either ignite motivation on the part of their staff or else face resistance; it will be peppered with red flags warning them that something may be about to go wrong. What are likely difficulties the managers will encounter when wanting to meet deadlines?[7] What aspects of the business will be most debilitating? Will the staff do what they are supposed to do, and do so every day, from start to finish?

How can a project targeting a wide range of end users see the light of day when marketing and project managers cannot come up with the right answers and solutions to these questions? They are twins in this business model. When innovation works out, the web provides help instead of hindrance and groups cooperate honestly and consistently; when commitment prevails over resistance, the twins can win. However, not too far away, a **utility drawback** always looms; time may be wasted and costs incurred may inflate, compromising quality.

Project managers rely on marketers to ensure that promoters, investors, and end users understand the project (one free of utility drawback), while marketers rely on project managers to ensure their promised delivery of a much-needed, innovative product sounds true. What would happen if that were not the case? Well, a delay in the vision phase of the project occurs, just when marketers and project managers outline the parameters and weigh in on project investments yet to come, and the delay acts much like a compounded interest charge on the start-up money. Planning, which prepares for mobilizing the people, the managers, and the resources, as well as deploying all the means and forces of production, finally giving birth to the anticipated product, will also suffer. In fact, along every step of the way, a poor match between marketers and project managers may hamper the project and its plans for marketing, perhaps even in a dreadful inflationary manner.

---

[7] For the full newspaper snapshot, see www.nytimes.com/2002/03/24/world/under-pressure-chinese-newspaper-pulls-expose-on-a-charity.html, Accessed January 20, 2019.

While project managers have often trained well in evaluating risks, they are not necessarily acquainted with **points of vulnerability** (POVs), and neither are marketers for that matter. POVs are internal to the system; they are inside the company handling the project. **Risks**, meanwhile, are always outside the control of the company handling the project. Managers can only try to anticipate and prepare for these risks, but they cannot control them. Risks come in eight forms:

1. Environmental,
2. Financial,
3. Legal,
4. Marketing,
5. Organizational (organizations outside the company handling the project),
6. Political,
7. Sociocultural (e.g., value clashes within the community), and, of course,
8. Technological.

Vulnerabilities do not refer to these areas; they refer to the human aspect of the business. A lifeguard does not see risks when facing a lake because he feels invulnerable. On the other hand, a child, who has not yet learned how to swim, is vulnerable and may more clearly see or intuitively sense the associated risks. Risks and vulnerabilities complete one another. Marketers and project managers cannot control the risks, but they can certainly detect and anticipate, as well as palliate, the vulnerabilities. Every marketing and all project efforts harbor POVs, if only because they involve people who must work together to achieve a goal.

Marketing projects is precisely about detecting the vulnerabilities along with the risks, strengths, and opportunities. Marketers and project managers do this prior to engaging too many resources. In short, they ask themselves: Given our internal vulnerabilities, given the risks we face, given the strengths our company has, and, finally, given the appealing market opportunity, is our project feasible? Is it possible to make it responsive to end users' needs? This is not a mundane question one can set aside, quite the contrary. This question is vital to the success of every project, and answering it properly is the best way to foster commitment, to ensure maximal use of online information and other technologies, to build group cohesiveness, and to come up with a truly innovative product—one that will not fall into obsolescence any time soon.

To explain what POVs are—an essential concept in marketing project analysis—we like to cite Victor Hugo, who describes a loose cannon aboard a war vessel:

> You cannot kill it, it is dead by nature. Yet, it has a life of its own. It has a sinister life that is injected into it from infinity. The floor, moved by dreadful waves and hellacious winds, is causing it to bounce around and lurch without coherence.[8]

---

[8] Our translation.

As one can guess, managers cannot, and must not, ignore POVs; yet the latter are often hidden. Our experience tells us they are present, at a minimum level of 4%–7% in any project. They represent a source of clear and present danger during any phase of the project, and lay inside its four constituting components: plan, processes, people, and power. They emerge because of poor measurements and poor metrics. They generate costs, create uncertainty, and breed conflicts; employees settle on disagreements and face-offs instead of finding solutions—they resist! Unrealistic and poorly articulated plans are typical of POVs, and marketers as well as project managers must deal with these; they must avoid diffused measures of success, inadequate change processes, incompetent managers (sometimes their own), and a lack of safeguards, all of these to prevent the project from derailing. Any experienced manager familiar with these hazards knows they cause havoc and compromise the project and its marketing's critical paths. In short, they fuel the project's utility drawback and render the project useless.

Both marketers and project managers are concerned with efficiency and efficacy. **Efficiency** entails maximizing the use of resources and eradicating undesirable wastes and errors; marketers want to sell products as perfectly fit to meet the customers' needs as possible. **Efficacy** refers to the ability to produce what the suppliers want to produce, that is, the capacity to complete the project according to the preset norms of quality, on time, and within budget. In marketing terms, this simply means offering exactly what the customers expect. As seen, marketers rely on project managers; but the latter also depend on the former, who provide the logic and reason justifying the project. Hence, both types of contemporary managers must imperatively coordinate their efforts. In fact, more and more jobs call for marketing project managers indeed.

To achieve success, both types of managers—marketing and project—rely on proper planning, carefully organized plans, committed stakeholders, and clear lines of authority that exercise proper controls. Among the stakeholders are the regulators (government, professional associations, etc.), the financial and material suppliers (subcontractors, clients, and end users), and finally, the unwanted individuals (groups or organizations, such as competitors or pressure groups, who have a keen interest in seeing the project—and the new product or deliverable— fail). Group cohesion within the project, including within the marketing group, is essential. For one, it tends to reduce turnover and encourages more participation in group activities,[9] something increasingly important in a world where virtual teams are gaining ground. Yet, no project is free of hassle. Even when things go well, people can use cooperation malevolently[10] and hostile behaviors can appear, such as when one individual feels others are better treated, or feels somehow deprived of some advantages or benefits compared to others and hence exerts revenge.[11]

---

[9] Carron, Widmeyer, and Brawley (1988).
[10] Hammer and Yukl (1977).
[11] Guimond and Dambrun (2002).

In one way or another, the main causes of failure in organizations relate to people. Some of the top causes attributed by managers are ambiguous goals, conflicts, insufficient allocation of or access to resources, lack of commitment, mediocre planning, inefficiencies in change management, unrealistic calendar of tasks and activities, and poor communication. All these problems affect both the marketing and project teams; this is something they have in common, not only in success but also in failure.

Sometimes, citizens see project managers as cold-minded engineers or mere coordinators who cannot empathize with customers or end users, while they perceive marketers as blunt liars who abuse naive customers. In most cases, these perceptions need correction. In fact, there are some out-of-track business sectors where marketing and project management are required. To give a few examples, we can think of the marketing of violent projects as done by terrorist groups or gangs, prostitution and pimping, pedophilia, predatory financial activities, illicit drug distribution in both national and international markets, as well as human and organ trafficking. Even in such environments, marketers and project managers espouse various objectives, with a fast-earned hefty profit being an obvious goal. Even bike gangs will go as far as trying to improve their social image or make their activities look appealing and cool to young, rebellious males. Hence, marketing and project management is not only about fulfilling needs, but also about acquiring market shares, beating competition, improving one's image, and generating profits (of course). Figure I.2 illustrates possible objectives that unite marketers and project managers in any economic sector, whether legal or illegal.

In our own ways, we all participate in fulfilling our needs, expanding our social networks, improving our image, seeking good incomes, and standing above the crowd in areas we cherish most—including love. We all market ourselves when we

Figure I.2 Possible objectives that unite marketing and project managers in any legal or illegal economic sector.

Note: Marketers and project managers share common challenges, even in economic sectors one would not necessarily think of as a sector of interest.

seek employment or a romantic partner; we all engage in projects, such as getting married or renovating our house. To put ourselves in the best light, we promote our projects and market them the best we can. We augment the perceived value we hope others see in our endeavors, be it a group vacation in a part of the world never before explored or forming a music trio. Marketers and project managers are not any different; they work with what marketers call **augmented products** because projects that see the creation of new products improve our way of life and, more often than not, include not only the product itself but also some services, guarantees, and even an emotional experience. To achieve this, they (and we as well) dig into our interpersonal and professional competencies; we rely on our knowledge, and we evaluate our environments.

GENERAL INTRODUCTION, CLASS EXERCISE #1:

Give known examples related to the above-mentioned list of common goals between marketers and project managers. Most of the readers should be able to relate in one way or another.

PROPOSED QUESTIONS FOR DEBATE:
1. Why do we conceive projects?
2. Why do we engage in projects?
3. What can project managers do to market their projects?

No matter the project, be it painting the Mona Lisa, building the Egyptian pyramids or the Great Wall of China, or even drafting a new law, we all work with the same base. Marketers and project managers must carefully coordinate their efforts as products grow increasingly complex, potentially fulfilling more than one need at once. A car, for example, is a response to a variety of needs: from moving between Point A to Point B to showing off our fortune or publicly expressing our social status.

When we question people about what unites marketing and project management, responses may take a while to come out. Then, words like influence, needs, and risks arise. When we ask marketers and project managers how they think they can work together using just one word, what comes out are words like cooperation, trust, and clarity of goals, among others. Table I.1[12] provides a summary of key words and concepts found in two projects analyzed by academics. As we will see throughout this book, the same concepts appear in a number of projects, and many project managers (as well as marketing managers who work with teams) will no doubt identify.[13]

---

[12] Marketing for the vast majority of people.
[13] Van Marrewijk et al. (2007), interpreted by us.

Table I.1 Example of Two Projects and Their Key Words/Concepts

Estimated Key Success Factors (KSFs)	Estimated Key Failure Factors (KFFs)
The Environ Project (Singapore)[a]	
Well-defined mandate	Lack of clarity in goals
Flexibility	Obscure scope
PPP[b] collaboration	Loss of control, or possibility thereof
Client support	Unexpected political changes
Values of innovation	Lack of stakeholders' commitment
Centralized management	Shortage of expertise
Traditional culture	Inefficient role distribution
The NSTP Project (Philippines)[c]	
Clearly defined mandate	Overly tight deadlines
Collaboration	Slow decision-making
Community involvement	Conflicting interests
Risk sharing	Us-against-them mentality
Good communication	—

[a] See http://environprojects.en.hisupplier.com/.
[b] PPP: Private-Public Partnerships. Vvan Marrewijk et al. (2008).
[c] Seehttps://fr.scribd.com/doc/5924351/The-Nstp-Project-Cesar-Community-Extension-Services-through-Action-and-Research.

Even though these examples pertain to project teams, they could easily apply to marketing teams. Marketing and project managers face the same reality. Thus, it appears wise to understand this reality from two sides: the back office so to speak, which is the project management office, and the front office, which is the marketing one.

Can we attribute the meager level of success most projects experience to faulty coordination between marketing and project managers? (Only 30% of projects meet their deadlines, respect their preset quality standards, and fall within the preset budget.[14]) Not at all! Problems occur at all levels, both in the back and in the front offices. Certainly, harmonizing the efforts of the back- and front-office teams may reduce POVs, points where the system may fail or altogether crash. This is why

---

[14] Dalal (2012); www.standishgroup.com, Doloi (2011), Buchanan (1991), and Mackenzie (2011).

we perform feasibility analyses, and why marketing feasibility addresses the question of whether a project can realistically meet customers' needs. Anything that can potentially wash out what the answer should be—a firm "yes"—must be carefully examined, be it poor loss management or internal conflicts.

Indeed, many projects fail because their promoters did not scrutinize their potential internal weak spots, or POVs, ahead of time, such as misjudging the chosen team members' competencies and their ability to work together—two fundamental aspects of any business venture.[15] At the end of the day, humans produce errors, and machines only make mistakes inasmuch as they are designed, built, programmed, and operated by humans. Weak projects and poorly designed products consume management's time, require frequent adjustments, draw resources away from profitable ventures, and cast a negative shadow upon the team.

I.1 What to Expect

This book intends to bridge the gap between marketing and project management. We will discuss their respective fields of expertise, their commonalities, and their differences. We will show how **dominant strategies** (DSs) guide project managers in their efforts to supply the deliverable within the specific constraints of time, cost, and quality.[16] To our knowledge, the "bible" of project management (PMBOK) does not address this gap, neither do most marketing books. Both marketers and project managers start from the same premise; there is a need, or more generally an opportunity (e.g., to expand market share, to make a profit), that requires a response, an action. Both will develop a strategy and use their own tools to implement their plan; both intend strategically, functionally, and operationally to reach their main goal: success, for which each will set its own standards. For marketers, this may be to increase usage (or attendance) or to put into the market a wider variety of products, or to offer a customer-friendly service. For project managers, the challenge is to meet the deadline, operate within the set budget, and respect the preset norms of quality. It is fair to say that both marketers and project managers would be pleased and congratulate each other if their joint efforts led to customers' satisfaction, a higher perceived value (and/or brand-image value) on the part

---

[15] Kloppenborg and Opfer (2002).

[16] Typically, project management has three accounts dedicated to risk management: a provisional account for known risks (addressed in the **DS**), a provisional account for the anticipated yet not necessarily known risks (addressed in the **contingency strategy),** and a provisional account for unanticipated and unknown risks. PMBOK 5 refers to "preventive actions (as) an intentional activity that ensures the future performance of the project work is aligned with the project management plan (...)" (p. 81). Finally, risk managers refer to risks that are unknown and unanticipated (**short strategy**) and typically have a provisional account to that effect. Marketing managers do not generally use such a system.

of these customers compared to both direct and indirect competition, as well as to customers' loyalty (if applicable) and profitability.

This book offers a general view of both marketing and project management. It does not intend to go into detail, but instead aims to provide new insights. Often, we challenge preset ideas we have found to actually be ineffective in the real-world market. We adapt some theoretical elements to fit better the requirements inherent to merging marketing and project management. For example, we update Maslow's rather static theory of needs, extensively referred to in marketing theory, by combining it with the concept of utility found in economics.[17]

Chapters 1 and 2 cover the general marketing concepts with which project managers should be familiar, and professors or tutors can use it for basic marketing courses. Chapter 2 focuses on marketing management, while Chapter 3 discusses the nature of projects. Chapter 4 explains what project feasibility is all about; something only a few books do, yet something that is essential in creating an innovative world. In Chapter 5, we discuss in more detail what marketing and project management have in common and what challenges their commonalities. Chapter 5 assumes that the members of the team that start a new project know each other; for example, they may all be members of the same organization. This contrasts with Chapter 4, which took for granted that the project started afresh. However, in real life and especially among contemporary companies, team members who have worked together on past projects commence new projects every day. The fundamentals of feasibility analysis remain but the analysts have a better knowledge of the work atmosphere ahead of time, which reinforces their reports. Chapter 6 provides testimonies we have collected over the years, including recently, which we believe shed a pragmatic light on the theory we cover. We conclude and offer a snapshot at various exercises (Appendix 1) meant to train readers—business people and students alike—to think in terms of feasibility and, more particularly, in terms of marketing projects. (These exercises are offered in full in our seminars.)

1.2 Unique Features

We decided to make this book as friendly and as practical as possible, without overloading it with academic references. We take a strong stand against some preconceived ideas and concepts that, in our view, disfigure marketing and project management when combined. These ideas and concepts may stand on their own in the silos of marketing and project management, but need adjustment when merged. When we take a strong stand and affirm an error exists in the literature, we identify

---

[17] Although it is used in marketing, no research to our knowledge has ever confirmed it. In fact, it is better to consider it as a working model; for example, needs are not always necessarily in the order stated by Maslow's rigid model nor are they necessarily sequential. However, for the sake of our argumentation, we adapt the model and give it some flexibility.

it with quotation marks, as follows: "arguable error." In Appendix 3, we list the "arguable errors" identified in this book; we do not claim to hold the truth but welcome debate. Our interest is in ensuring marketing theory has pragmatic applications. This way, readers are aware we are expressing our subjective viewpoints. We do not list the books or authors we feel have made such errors by due respect; most often, these are common mistakes found in many different books. As an example, some authors describe the purchasing process without ever allowing the actual purchase to take place. It is therefore a purchasing process without purchase, which, to us, does not make sense.

We do not expand at length on core concepts; often, we simply give a base definition and a short argument, with the purpose of always fostering mutual understanding of the two disciplines: marketing and project management. There are countless books on either subject, but few, to our knowledge, exist on marketing projects. We make a rather heavy use of figures and tables to serve as mnemonic devices; our experience taught us that, currently, many readers skim written information and prefer to capture the intended meaning by way of images. The reader should take time to consult the figures and tables to maximize understanding of the material presented.

Within each chapter, we also provide suggestions for exercises readers can do on their own, or with a professor or mentor/tutors in workshops, seminars or classes. Generally, these exercises allow the participants to relate more personally to the theory we cover. They are often a great way to initiate discussions among the business trainees and students, and have been found to increase motivation in further learning. For that purpose, we also raise questions aimed at triggering debates. We insert the occasional tutorial note or footnote to guide the trainer's or professor's efforts. We provide a mind teaser section at the end of the general introduction and of the subsequent chapters. Occasionally, we insert short sections called rule of thumb, which can serve as simple reminders or managerial tools.

Our goal is not to turn the readers—either business people or university students—into experts in marketing and/or project management. What we want to achieve, rather, is the development of the ability to perform a cold diagnosis of the state of the business without going into great depth. We call this **upstream thinking,** or if one wishes, thinking before thinking; or better yet, thinking before thinking before acting. This will all become clearer as we evolve together through the upcoming chapters.

This book is merely an eye-opener that attempts to promote sound managerial habits. We provide a great number of tools readers can (and should) use, some of which admittedly require a fair bit of practice. During our seminars, we offer exercises that do not pretend to be all purely marketing, nor purely project management, nor a perfect fit for marketing and project management. Instead, we based our exercises on real cases, allowing readers to train their brains to think in a way that marries marketing and project management. Our experience in testing these exercises has given us confidence they are effective in helping both trainees and students view business challenges in a new light. We have also noticed that those who work together, and apply the principles and tools provided in this book, work better in teams down the road, especially when they address other managerial problems. We recommend readers/seminars' participants do the exercises meticulously: They are a great way, we think, to prepare business people and university students alike in upstream thinking. Only through doing these exercises can one gain a sense of what upstream thinking is all about, and, with it, marketing projects.

As for the quizzes provided at the end of each chapter, we endeavor to make the questions more difficult for pedagogical purposes as we progress along in this book.

I.3 Mind Teasers

Readers may use the mind teasers as questions in preparation for an examination or quiz.

1. List
   a. five possible objectives that unite marketing and project managers and
   b. four contemporary challenges shared by marketers and project managers.
2. True or False?
   a. POVs are points where the system may fail or crash altogether.
   b. Project capability refers to "the appropriate knowledge, experience, and skills necessary to perform pre-bid, bid, project, and post-project activities."[18]
   c. The goal of DSs is to guide project managers in their efforts to deliver the final project within the specific constraints of time, cost, and norms of quality.
   d. Upstream thinking is thinking before thinking before acting.
   e. Utility drawback is about starting a project all over again.

---

[18] Davies and Brady (2000, p. 62).

Chapter 1

What Is Marketing?

An eco-friendly building in Strasbourg, France. The surrounding community envisions an entire area of such eco-friendly buildings.

1.1 Introduction

This chapter briefly reviews the main concepts pertaining to the marketing field and that are relevant for project managers. We do not pretend to cover everything there is to know about marketing; in fact, we even challenge some preconceived ideas, but do so for the benefit of harmonizing the roles of the two experts: the marketing and the project managers. We resort to a short-and-sweet approach, as our goal is to briefly present the topics and not delve into them too deeply.

The learning objectives of this chapter are to understand what marketing is, to recognize marketing activities and concepts in daily life, and to make the link between marketing and project management.

1.2 What Is Marketing in the Context of Project Feasibility?[1]

According to the American Marketing Association (AMA), marketing is "the activity, set of institutions, and processes for creating, communicating, delivering, and exchanging offerings that have value for customers, clients, partners, and society at large."[2] Some books state that marketing is about the creation of value. We believe this is "arguable," at least from a project management perspective. Operating a machine is also about creating value, for that matter. In fact, many products people put in the market have little or no value at all, or even have a negative value. What about blinds containing lead that poison kids when put in their bedrooms? What about opioids becoming a considerable problem in North America? It is hard to argue these products create value, yet marketers still put them into the market.

Simple definition: **Marketing** is to put into the market.

Walking is to walk. Marketing is to put into the market.[3] It is as simple as that. Books and theories that shy from this simple explanation lead readers in the wrong direction. Projects' promoters want to see their proposals accepted by investors (funders). Thus, they need to find a way to put products in the market of investors. Once completed, and thus having become an operation, managers must put the deliverable into the market. It has to reach its intended end users.

---

[1] We interpret marketing concepts within the realm of project feasibility. Some of our statements may differ from the standard marketing theory.

[2] See www.ama.org/the-definition-of-marketing/.

[3] The word "market" has a Latin origin. Note that we endeavor to keep all definitions in a format as simple and as easily memorizable as possible.

On the other hand, marketers are usually keenly aware of latent, or unconscious needs, as well as those developing and emerging, and can provide project promoters and managers with exciting ideas. In the end, both marketing and project managers long to deliver value to the funders and future end users of the projects; they choose what value they wish to offer, work on building it, and provide and communicate it.

1.3 Who Are the Marketing Experts?

We divide marketing into two areas of expertise: the analytical side and the creative side. The analysts gage the market, trying to identify trends, **hidden needs**, and competitors' weaknesses, and they constantly attempt to better capture buyers' and consumers' behaviors. These analysts thrive on statistics, love mathematics, and spend many hours scrutinizing databases such as those provided by, say, Euromonitor[4] or AC Nielson.[5] On the other hand, some marketing experts excel in creating ads, logos, and slogans (sometimes called mantras), or else in working hand in hand with salespeople to create innovative promotional programs. Jobs include Vice President (VP) of marketing (or, often, VP of sales and marketing), Director of marketing, etc. Pay varies widely, but a good career in marketing can be quite rewarding financially.

The skills required for the marketing field include a strong mathematical and analytical sense (to set prices and analyze data), sensitivity to market agents (especially the buyers[6]), imagination, and initiative.[7]

Marketing experts occupy an important role within organizations. Unfortunately, our experience has shown us that employers often blame them for market upheavals, even though they have no control over them! Sales are going down, so top managers must find a culprit.

Marketing experts work closely with research and development (R&D), especially when it comes to projects—the mechanisms used to conceive, test, and eventually create and "operationalize" new products to sell. Increasingly, sound knowledge in project management is becoming necessary for marketing experts, just as marketing knowledge enhances project managers' ability to realize their ideas.

---

[4] See www.euromonitor.com/.

[5] See www.nielsen.com/fr/fr.html; for further examples, see also opinionresearch.com, google.com/intl/fr/analytics, stat.gouv.qc.ca, marketingpower.com, dialog.com, or lexisnexis.com.

[6] This is important in international marketing.

[7] Individuals interested in such a career should favor companies with solid reputations; those that have adequate personnel and stable financials provide sound management, offer something of value to the market, and are integrated within their communities. These conditions, when filled, make any marketing job that much easier and that much more promising.

1.4 Marketing for the Vast Majority

The vast majority of people have an ambiguous understanding of what marketing is; many have preconceived (and often negative) ideas, or else do not grasp its full scope. In fact, the main themes usual marketing academic journals cover provide a succinct list of those anyone can think of, including brand management, consumer behavior,[8] culture, ethics, innovation and product development, international marketing, publicity, sales and customer relations, social networks including an Internet presence, strategy and decision-making, and viral marketing.

Marketers resort to market analyses, while sellers focus on the interaction between suppliers and buyers. Technically, marketing and selling go hand in hand. Project managers or promoters market and sell their projects: They devise strategies to put them into the market, and then walk the talk and meet with their chosen audience—the buyers and/or end users.

One of the most surprising things about marketing science is that it focuses a lot on consumer behaviors, but barely those of the sellers, except in the context of sales. However, marketers often work closely with salespeople and must think about their marketing strategies in terms of meeting both the customers' and the sellers' needs. Project promoters are selling agents in their own right, and they too must align their behaviors with those of the intended customers/end users. Therefore, we must view marketing as an all-encompassing effort whereby marketers design a strategy to put something into the market while considering the role of both sellers and buyers.

CHAPTER 1, CLASS EXERCISE #1:

Define your perception of marketing. Common words tend to include "selling," "a bunch of crooks," "strategy," "publicity," "promotion," "needs," "consumers," and so forth.[9]

PROPOSED QUESTIONS FOR DEBATE[10]:

1. Do marketers abuse consumers' naivety?
2. Do marketers create artificial needs?
3. Do we need portable phones?

---

[8] Consumer behavior is really about the behaviors of those who actually consume the products. Hence, it technically only looks at a small portion of the overall spectrum of behaviors, particularly when it comes to consumption. For various reasons, many buyers do not actually "consume" the products they buy.

[9] Many people have a negative view of marketing because they associate some of the sellers' shenanigans with marketing tricks. This is not always false, but marketing is not selling—and tricksters exist in any profession.

[10] Tutorial note: Engaging in debates with students, or participants, is a great way to get them excited about the topic and realize what preconceived ideas they host, which we try to eliminate when doing upstream thinking in the marketing feasibility of projects.

1.5 What Do We Put in the Market?

We have now briefly defined marketing, so the next question is: What do we put into the market? Most marketing books cite products and services, but this is incomplete. We put five things in the market: products, services, ideas or messages, experiences, and projects (see Figure 1.1).

Products are tangible, standardized, and durable to a certain degree. (They can range from very durable to non-durable, such as perishable foods, meaning they have a shelf life.) We can count them in units (grams, inches, etc.). Services are intangible, inseparable from the provider, ephemeral, and change every time we use them (e.g., we never get exactly the same haircut).[11] Some marketing books argue that products and services are one and the same but, as one can judge by their individual characteristics, they are clearly two very different things. (We consider this an "arguable error.") Products and services extend on a scale that starts with such durable goods as clothes, jewelry, furniture, and houses, and stretches toward services such as haircuts, childcare, and medical diagnosis. Marketers also classify goods by the type of buying experience: spontaneous (a sudden craving for an ice-cream cone when it is hot), common (purchases such as a BIC pen), and specialty (products such as a wedding dress). Marketers also segregate products as to whether they are intended for consumers or industrial outlets, by the type of retail outlet (e.g., large, medium, small), or by the product's adaptation to specific needs (mass markets *versus* customization).[12] Projects are generally well thought of, especially if they require a feasibility analysis.

Figure 1.1   Five things we can put into the market.

Note: A market includes products, services, ideas and messages, experiences (or experiential marketing), and, increasingly, projects. Historically, societies improve their offerings as they move from products to projects.

---

[11] Marketers often use a measuring tool called SERVQUAL to assess the quality of service. It includes measurements on reliability, empathy, responsiveness, availability of "tangibles" (e.g., a computer room at a hotel), and assurance (confidence the employees generate).

[12] See the section on segmentation.

Political campaigns are a perfect example of ideas or messages. While products attract the conative aspect of human beings, and the service their emotional side, ideas and messages generally appeal to their cognitive self. More and more sophisticated products provide a mixture of these elements, and the more they do, the more memorable they are and hence they develop into an experience. For example, Walt Disney provides memorable experiences for millions of children.

We can market projects because they are a mixture of all the above. Projects are an ephemeral experience because projects have a beginning and an end, after which they become an operation. A project is thus a holistic experience that generally attempts to develop a product or service people will use—or both, within the frame of an overall experience. We call that product or service a **deliverable**.

1.6 Why Do We Market Products, Services, Ideas, Experiences, or Projects?

Marketing theory talks a lot about needs. There are different types of needs and different levels of neediness. Plenty of theories exist on the subject, yet the conventional marketing theory often falls a bit short of other facets of motivation. There are four, in fact: desires or wants (e.g., a longing to become wealthy), problems (especially true in mechanical structures, where wear and tear occur), sources of discomfort (e.g., a sickness that requires treatment, or a dirty diaper on screaming babies), and finally, needs *per se*.

Experts can hardly prepare a sound marketing plan without understanding what type of motivation they must address. Needs are to the customers what opportunity is to the sellers (see Figure 1.2). Managers conceive and materialize projects to fulfill a desire, solve a problem, soothe a source of discomfort, or fulfill a need. The economic implications of such differentiations are huge; needs must be dealt with first, before mere desires. As such, a project will suffer more from the

Figure 1.2   The consumers' motivation is the sellers' opportunity.

Note: Marketing and project managers can view a project as two sides of the same coin: motivation for the user and opportunity for the manager.

pressure to complete it on time depending on the source of motivation that ignited its undertaking in the first place. This very concept skips the attention of some marketing and project managers, yet it is relevant in all aspects of modern business.

1.7 Where Does Marketing Theory Come From?

The science of marketing is relatively recent, just as is that of project management, even though humans have done both (although not formally) since the early days of human evolution. Personalities include Butler, DeBower, and Jones,[13] who are credited with the first official use of the term "marketing" in 1914; Jerome McCarthy, who introduced the 4Ps of marketing—product, price, place, and promotion—in the 1960s; and Everett Rogers,[14] who talked about the diffusion of innovation back in 1962. As we will see, innovation is a core variable of the definition of projects. Hence, early on, market and project management have had commonalities.

Academic journals emerged as the marketing theory evolved, with Americans leading the way with the creation of the *Journal of Marketing* (first published in 1936) and the *Journal of Marketing Research* (1964). In 1964 came the *European Journal of Marketing; Recherche et Application en Marketing* (RAM) followed in France in 1984. Other events have marked the rise and growth of the marketing science, including

1. The Sherman Act of 1890 (United States), which aimed at curbing predatory pricing (or, pricing devised to strip competition from the market),
2. The creation of the American Marketing Society (AMS) in the United States in 1931, which was the first marketing association (the equivalent in France was the Association française du marketing, which opened in 1984),[15]
3. The voting of a law in 1956 against subliminal advertising in the United States, and
4. The use of advanced algorithms in publicity placements on Internet services (e.g., Google) and social networks.

World War II saw the development of propaganda, especially under the spell of Joseph Goebbels, a tactic that Charles de Gaulle used in his own way starting on June 18, 1940 (notably through the British Broadcasting Corporation [BBC]).[16]

---

[13] Butler, DeBower, and Jones (1914).
[14] Rogers (1962).
[15] There are many marketing-related organizations, such as the Canadian Association of Importers and Exporters or the Print Measurement Bureau.
[16] In his memoirs, de Gaulle mentioned how he warned the Germans over the radio that French resistance activists caught and executed by the Germans would see an equivalent number of Germans executed. Mass communication commenced henceforth.

Table 1.1   The Four Main Eras of Marketing

Name	Description	Approximate Decades
Production	Engaging in large-scale production and mass appeal (i.e., Ford's Model T)	1870–1920
Sales	Selling at all costs	1920–1960
Marketing	Understanding the market	1960–1980
Client focus	Empathizing with the client (or, at times, considering the client as "king")	1980–today

Hollywood, of course, and in particular Metro-Goldwyn-Mayer Studios, Inc. (MGM), also resorted to propaganda and distributed movies throughout the United States to convince Americans to join the war effort.

Historians divide the historical evolution of marketing science into four main areas, as given in Table 1.1.

As can be seen, marketing paid initially little attention to the consumers' personality, preferences, and needs. However, with time, and the influence of some European academics, marketing developed from an anonymous approach to customers to a personalized one, assisted by the advent of computers that allowed for accumulating and analyzing massive banks of data on purchasing habits. Perhaps projects evolved in a similar way, as engineers rely more and more on advanced technical tools and as they better understand the end users' needs, preferences, and habits. Both marketing and project managers have learned, over the decades, to become more sensitive to their consumer base.

Experts have also developed different forms of marketing over the decades, to which many readers have been exposed. We can think of viral marketing, green marketing (focusing on the social tendency to promote a greener planet), database marketing (tracking consumers' habits), customer relationship management (CRM; a trend that puts customers at the forefront of economic activity), and so forth. Project management does not have such various kinds of formats: It is monochromatic from that point of view. Occasionally, project managers are exposed to marketing groups (and even to lobbyists), which interfere with their projects. The marketing of projects must thus be sensitive to potential of disruptions, which is something we will see when we discuss project feasibility (Chapter 4).

1.8 Where Does Marketing Take Place?

Marketing takes place around the globe. Most governments provide critical information on their websites about their markets, as do a number of other sources such as the World Trade Organization. This does not mean national experts exercise marketing with the same practical understanding; typically, Americans are more advanced.

Publicity is one aspect of marketing the majority of people face on an ongoing basis, at times receiving thousands of messages daily. In 2018, companies in the United States spent 13.3 billion USD in advertising for a population of 310 million people (China: 11.6 for 1,300, Japan: 2 for 127).[17]

Marketing science borrows a lot of its terminology and concepts from other domains, such as the military, as well as psychology, sociology, and economics. The military domain has influenced marketing, indeed. We call a favored market a target market; many of the strategies employed in market penetration use military terms and concepts, such as marketing "encirclements" and "campaigns."

Marketing also uses many psychological concepts; for example, in the 1970s, publicists used fear to entice consumers to buy their products.[18] Along that vein, marketing also uses concepts pertaining to sociology, especially because marketers are much interested in group phenomena.

Marketing science deals with sellers and buyers, and, just like economics, supply and demand as well as satisfaction, utility, product substitutes, and risk. In fact, both fields often view consumers' behaviors from their own biased viewpoints, which are in fact complementary. As an example, marketing experts talk of desires and economists of preferences.

Classical economists assume that consumers are rational, that their trading activities occur without regard to time limits, and that they are well informed. Marketers are generally more nuanced: Some consumers are not desirable at all (because they misbehave, complain too much, cheat or steal, etc.), some are clueless as to exactly what they need or want, and some are misinformed (a fact some astute marketers use to their advantage). Consumers do indeed get impatient when they wait too long in line, or angry when what they buy does not deliver.

Project managers are thus closer, conceptually, to marketing science than to the economy because all projects are bound by time. We will see further along how important it is for project managers to understand consumers/end users in a much broader light than the one used by economists.

1.9 What Defines a Market?

There are different ways of defining a market, which we briefly cover below. Broadly speaking, a market is an area (real or virtual) where agents trade goods (e.g., products, services, money). Classical economy talks of perfect markets. These are markets where consumers always find what they are looking for at the price

---

[17] Source: ZenithOptimedia, groupe Publicis, IMF.
[18] This is not always done to the best interests of consumers, as many people will reproach the marketing experts.

they wish. If a supplier is not able to offer the needed or desired product, then another supplier seizes the opportunity. In reality, perfect markets do not exist; yet they provide a model that experts use to analyze consumer behavior in a simplistic format. Many markets are actually composed of large companies, which together control a large percentage of the market in the business sector with which they deal. This market is an **oligopoly** and numerous examples exist, including in the United States (e.g., the accounting firms known as the Big Four). A short form of oligopoly is a **duopoly**, where only two players act in the market.

On the other hand, **monopolistic** markets are markets where only one player provides the products or services coveted by consumers. In such markets, of course, there is no competition so the unique supplier can set the price it wishes. Many advanced countries have anti-monopoly regulations; competition is good, as consumers end up with more choice and suppliers generally treat them better out of fear of losing their business. Generally, economists also consider that monopolies don't make the most efficient use of resources. Some marketing books posit that markets are a territory where companies compete, but in a monopolistic market, there is hardly any competition. (We consider this an "arguable error.")

Projects take place in all kinds of markets, be it an oligopoly or a near-perfect one. This affects not only how managers source the projects, but also how they manage them. Experienced project managers are fully aware of the subtleties of doing business in countries where local habits vary and economic powers are peculiar.

Simple definition: A **market** is a real or virtual area where sellers and buyers exchange goods (e.g., products, services, projects) and currencies.

We can also view markets according to their colors. This seems weird, but it is true, and few people actually know about it, but project managers face such markets in certain countries, for example, those where corruption is high.[19] To illustrate, a **regular market** (let us label it as **white**) is a market that obeys the laws of society: It is legal, and trading agents pay taxes and exercise their privileges in fairness to everyone—at least on paper. **Black markets** include prostitution (where it is illegal), drug dealing, human trafficking, and so forth. This represents a huge worldwide market that is worth many billions of dollars. Not to be mistaken, market agents working in black markets use and benefit from marketing science; they too have to devise strategies to put their products into the market (sometimes to re-channel their income and make it legal, like through money laundering) and to sell what they have (whether legally or not). Terrorist groups, too, form a market and their

---

[19] For corruption levels by country, see Transparency International.

agents use propaganda, among other tools. Therefore, projects exist in both the so-called white markets and black ones: Putting in place an international network of opium distribution serves as an example.[20]

Finally, there are **gray markets**. Gray markets are not as well known, but they do exist. They are composed of products that contain defects and which producers cannot sell through the regular channels because consumers would reject them, thus damaging their brand name. Marketing agents buy stocks from regular companies, which their quality control team[21] have rejected, and resell them in local venues (such as some 1-dollar or 1-euro stores), or else to countries where such products would be in demand regardless of their quality. Gray markets are at the intersection between white and black markets. In this book, we focus on white, or regular, markets and legal projects.

CHAPTER 1, CLASS EXERCISE #2:

List your sales and marketing experiences (buying a used car, registering on a website), and then ask yourselves what upset you most and what you admire most in marketing activities. This prepares you to become self-conscious of your biases, an ability needed in the realization of marketing feasibility of projects.

PROPOSED QUESTIONS FOR DEBATE:
1. Is marketing essential to economic activity?
2. Can marketing experts afford not to be at ease with basic mathematics or basic language skills?

Clients come in two main forms: existing and potential (or prospective). From a project management perspective, existing clients are the promoters, or fund providers (funders), and the companies they wish to please (e.g., a government having paid for a bridge). Potential clients are the end users to which the project's deliverable (e.g., a bridge) is intended, but who have yet to use it. Marketers wish to integrate potential customers and to secure existing customers' loyalty. However, as mentioned, not all clients are desirable. Marketers estimate that about 35% of clients truly meet their hopes and expectations: They bring in profit, give repeat business, and do not cause grief. The rest unnecessarily occupy salespeople's time; they complain, damage goods, show no loyalty, often return products unjustifiably, and so forth.

---

[20] The British Empire used such a tactic when it set out to conquer China in the 19th century.
[21] Typically, this is about 4%.

Walmart has set up a policy that plays on unwanted customers: the no-questions-asked return policy. Many customers buy products to use them over the weekend (e.g., an electric saw), only to return them the following Monday (after finishing their basement, for example). No questions asked. Walmart knows it does not lose money in the end because most customers, it found out, actually wander and stay in the store longer and end up buying more products, which those tricky customers will not return. The losses are absorbed by the fact that bad customers are dragged into the store to buy more products, which they cannot resist even if they don't need them. Not all retailers can afford such a tactic, but it works fine for Walmart. Small businesses would much rather send undesirable customers to competitors rather than exhaust their own limited resources.[22]

Another way of defining markets is by their consumer types: consumer (personal) or industrial markets. They buy/purchase products in specific ways (see Table 1.2).

As illustrated by the above discussion, at its heart, marketing fosters relationships with leads (who are possibly aware of the marketing offer, but are certainly not yet clients), prospects (would-be consumers), and consumers (whether industries or

Table 1.2 Characteristics of Consumers' and Industrial Buyers' Behaviors

Industrial[a]	Consumers
Not for personal consumption	Individual consumption
Purchases done by one other than the end user	Purchasers done by end users
Often involves a group of decision-makers	Decision taken by one or two individuals in general (e.g., a husband and wife)
Purchases respond to set rational and highly specific criteria	Flexibility in accommodating purchasing criteria
May require a tender	Generally, no tender
Associated with an inelastic demand (product is needed)	Substitutes generally acceptable, if available

[a] Marketing books tend to separate the industrial stakeholders as follows: the initiator (the one who notices the need), the user, the industrial buyer, the influencer (e.g., an engineer who knows best what specific characteristics a product must have), the decider (often, someone in the finance department), and the followers (who receive the products, store them, etc.). Medium and large projects, by nature, engage in substantial B2B trading activities.

---

[22] On the other end, some suppliers are also tricky. They will put on a nice face when attempting to sell their products or services, but as soon as they receive payment, their attitude takes a 180-degree turn and become non-responsive or abrasive altogether.

businesses [B2B], or people/consumers [B2C]). Prospects are aware of the marketing offer, pay attention to it, and may actually manifest some intent of further evaluation, buying the product in the end. They turn into customers when they actually buy said product.

Obviously, projects pertain more particularly to industrial purchasing. Thus, the relationship between project managers and their suppliers is highly technocratic. However, the relationship between these managers and the customers/end users gains from being interpersonal, rather than uniquely industrial.

To that effect, the three main marketing approaches to **customer relationships** encompass three different levels of influences. Table 1.3 summarizes these approaches.

This is also meaningful to project managers. More and more project developers and managers contact consumers/end users to share their experiences and to unearth novel ideas. Hence, our modern world sees an increasing tendency for cooperation between these stakeholders. The goal of achieving closer relationships ties in with the notion of experience seen earlier: Both types of managers envision consumers/end users living an exciting experience before, during, and after engaging in the project. This tends to secure loyalty and satisfaction.

Table 1.4 delves into more details as to what the three interactional levels between customers and suppliers involve.[23] To highlight the contrast, we have added a fourth level: the intimate one. This, of course, applies in different circumstances.

Projects are particularly sensitive to these various business levels; indeed, they typically are completed over a relatively small horizon and involve stakeholders who have previously had little opportunities to work together or engage in dilemmas, such as meeting a strict timeline while dealing with budget constraints and preset standards of quality. Hence, projects seem to be a more natural fit for interactional relationships; yet, most projects fail because of their human components as we shall see further along in this book. At some point during the project, trust, expected

Table 1.3   The Three Main Types of Relationships between Sellers and Consumers

Name	Description	School of Thought	Approximate Decades
Transactional	Contract based	American	1960–1980
Relational	Low/medium-level trust based	European	1980–2000
Interpersonal	High-level trust based	In development	2000–today

---

[23] Mesly (2010).

Table 1.4 The Three-Plus-One Interactional Levels

Characteristic	Transactional	Relational	Interpersonal	Intimate
Duration	Short	Long	Lengthy	Short to lengthy
Cooperation	With programmed exit	Adaptable	Assumed	Long term
Dynamic	Rigid (contractual)	Adaptative	Flexible	Ongoing compromise
Closeness	Poor (distant)	Minor personal components	Personal	Intimacy
Adaptation to others' needs	None or weak	Moderate	Intense	Revealing self
Emotional investment	Weak	Moderate	Strong	Exclusive relationship
Motivation	Egoistic	Egoistic or altruistic	Resolutely altruistic	Mutual
Object of trust	The organization	The individual	The person	The couple
Expected ethical commitment	Honesty and decency	Respect and equity	Empathy and sharing	Voluntary commitment
Predictability	High	Some imponderables	Moving	Common goal
Decisional aspect	A logic of calculated profits	A logic of goodwill and profit	Heuristic logic	Logic of compromise

ethical commitment, and the like exercise a large influence over stakeholders.[24] Hence, project managers can gain from understanding these types of relationship, extracted from marketing theory.[25]

1.10 Who Are the Other Market Agents?

We have so far seen two market agents: the sellers and the buyers. Sellers (or producers who will ultimately have to sell their products) are interested in production, efficiency, return on investments, and other such concerns. Consumers think in terms of satisfaction and need fulfillment (or, in economic terms, **utility**; that is, the capacity to fulfill a need). They may have as criteria such items as durability, brand reputation, functionality, and so on.

Quite a few marketing books tend to ignore that markets, in fact, are also composed of two other market agents who play a crucial role in the makeup of said markets: regulators and outsiders (see Figure 1.3). Some marketing books posit that other market agents include influencers and distributors. In fact, influencers are providers of opinions and distributors are suppliers as they sell their services. (We consider this an "arguable error.")

Figure 1.3 The four market agents.

Note: Sellers and buyers are a given. Sometimes, regulators are addressed and should certainly be considered in the context of project management. Disruptive outsiders should also always be of consideration in a feasibility analysis.

---

[24] In project management terms, we refer to this as PWP or work psychodynamics. There are six such strategic Ps in project management, with each addressing a particular analytical approach. See Chapter 4.

[25] There are various models discussing the relationships with customers, including customer relationship management, or CRM.

Regulators set the rules under which trading activities take place in the market, including rules of proper marketing conduct. We outline codes of ethics marketing organizations promote in Table 1.5.

Many laws ensure markets protect consumers.[26] As we all know, manufacturers use various symbols to protect consumers (toxic products, nuclear waste, etc.).

Table 1.5   Examples of Behaviors Expected From Marketing Experts

Organization	ACM[a]	ARIM[b]	APRM[c]	AMA[d]
Truth	•	•	—	•
Respect of laws	—	•	—	—
Professionalism	—	•	•	•
Competence	—	•	•	—
Respect of clients	•	•	—	•
Accuracy	—	•	•	—
Clarity	•	—	—	•
Trust from the public	—	—	•	—

Note: The dots highlight an item that is more particularly emphasized.

[a] Association canadienne du marketing—see https://online. the-cma.org/french/?WCE=C=47|K=225885, accessed April 26, 2014.
[b] Association de la recherche et de l'intelligence marketing— see http://mria-arim.ca/fr/a-propos-de-larim/normes/code-de-deontologie-des-membres, accessed April 26, 2019.
[c] Association professionnelle de recherche en marketing.
[d] AMA—see www.ama.org/Pages/default.aspx, accessed May 5, 2019.

---

[26] A few examples for the Canadian market are laws concerning alcohol (L.R.Q., Ch. P-9.1), author's rights (L.R.C. 1985, Ch. C-42), brand names (L.R.C. 1985, c. T-13), bankruptcy and insolvency (L.R.C. 1985, c. C-34), competition (L.R. 1985, C-34), consumer protection (L.R.Q., c. P-40.1), criminal code (L.R.C. 1985, c. C-46), dangerous products (L.R.C. 1985, Ch. H-3), foods (L.R.Q., Ch. 29–40), French language (Charte de la langue française. Loi 101, art. 1), lottery (L.R.Q., c. Ch. L-6), packaging (L.R. 1985, C-38), patents (L.R.C. 1985, Ch. P-4), personal information (L.R.Q., c. A-2.1), publicity (Code Canadien des normes de la publicité), radio and television communication (L.C. 1991, Ch. 11), security in the auto-mobile sector (L.R. 1985, C-24.2, r. 32), tobacco products (L.R.Q., Ch. T-0.01), and weights and measures (L.R.C. 1985, Ch. W-6).

Table 1.6 Examples of Organizations Defending the Interests and Rights of Consumers

Country	Year	Name
Canada	1947	Consumers' Association of Canada
France	1951	Union fédérale de la consommation
England	1957	Consumers' Association
World	1960	Consumers International
England	1963	National Federation of Consumer Groups
Canada	1964	Insurance Bureau of Canada
United States	1968	Consumer Federation of America
Japan	1969	Consumers Union of Japan
Québec	2004	Autorité des marchés financiers (AMF)

Furthermore, consumers from various countries benefit from the helping hand of many organizations. For examples, see Table 1.6.

These organizations intend to make consumers' rights better known, educate said consumers, and promote the enforcement of responsibility and the establishment of sanctions. Project managers are sensitive to these customers'/end users' rights, especially when these customers engage, in one way or the other, with the project.

CHAPTER 1, CLASS EXERCISE #3:

Search for similar laws and mechanisms that exist within your own country (if applicable).

PROPOSED QUESTIONS FOR DEBATE:

1. In what way do these laws, codes, and measures protect the markets of both buyers and consumers?
2. Are there tricks marketers can use to avoid some of these regulations?[27]

Marketing and project managers must obey those regulations pertaining to their own field (e.g., patent protection) and must follow proper standards (e.g., for

---

[27] This exercise has the potential of generating intense debate, helping students express some of the frustration they may have toward what they see as unfair marketing tricks. This prepares them to adopt, by contrast, the objective approach necessary for completion of the marketing feasibility of projects.

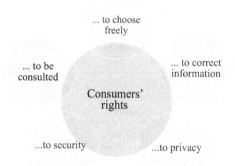

... to choose
freely

... to be
consulted

... to correct
information

Consumers'
rights

...to security

...to privacy

Figure 1.4 The five fundamental rights of consumers.

Note: Marketing and project managers should keep consumers' rights in mind at all times. Clients who work on projects will not forget them, whether consciously or unconsciously.

dangerous products). As such, they share some commonalities and will likely interact with respect to at least one of these items. The five rights of consumers apply very well to relationships between marketing and project managers. We show these **five rights** in Figure 1.4.

No doubt, marketing and project managers must guide their interactions by recognizing these five rights. Only then can they create a fruitful and trustful atmosphere. On the other hand, there are a number of reprehensible marketing practices (which project managers should avoid when developing their marketing expertise), some of which we list in Table 1.7.

Outsiders, the fourth market agent, are precisely those who escape, in one way or the other, the above-mentioned rules, codes, and regulations, and potentially resort to the above-mentioned deceitful behaviors.

Many projects are plagued, from the inside or the outside, by such demeanors (readers can easily find them by searching key words related to scandals in projects; they are countless!). While project management is not regulated, as are economic transactions and marketing activities, it remains that more and more project managers are becoming exposed to various market agents and must be cognizant of the good and the bad sides to inferred interactions.

1.11 Can Marketing Actions Be Misinterpreted?

Marketing experts, whether in the academic or the private sector, are not perfect. An inescapable example is that of Coca-Cola (Coke). In 1985, Coke set up a project to improve the formula of its star product and launch a new Coke. Coke had been

Table 1.7   Examples of Reprehensible Marketing Practices

Communication		
Dubious comparisons	Incomplete description	Visual distortions[a]
Exaggeration	Deceitful promises	False testimonies
Publicity targeting children	Shenanigans	False publicity
Subliminal publicity	—	—
Pricing		
Collusion	Price discrimination	Price fixing
Refusal to sell	False pretexts	Predatory pricing
Bait-and-switch tactics	Selling damaged or harmful products	Pyramidal schemes
Other Strategies		
No-obligation competition	Use of other people's suffering	Incitation to over-consumption
Building on consumers' impulsivity	Use of children	—

[a] For example, giving the impression the product comes in larger format than it actually is by way of packaging.

one of the top brand names in the world and was for many years; yet, its marketing gurus thought they could compete more effectively against Pepsi by reinventing the brand and producing an improved version of its classic formula (which comprises sugar, caffeine, carbonated water, and corn).[28] The move was a massive flop[29] and consumers became disillusioned, if not angry, at the change.[30] The project went kaput.

The new formula did not cause the market upheaval; in fact, it tasted better than the original one, as was confirmed by pretests[31] done with consumers by comparing the two. It stemmed from the fact that Coke did not respect the attachment consumers had to the brand name or its formula. It was treason, an attack on the way consumers identified with the brand.

Other renowned marketing mistakes concerns Ford, in the 1970s, naming their new car Nova in an effort to enter the Latin American market. (Nova means "star"

---

[28] Originally, Coke contained small amounts of cocaine, hence the name.

[29] Hirsley (1985).

[30] For the "From the Press" insert, see Chicago Tribune, Michael Hirsley (April 28, 1985).

[31] Project management sometimes uses the term **Proof of Concept** (POC).

in Italian but, unfortunately, means "don't go" in Spanish. Who wants to buy a car whose name says it is not going to make it from Point A to Point B?) Mazda's MR2 had somewhat of a dubious name for the French market, as its phonic rendition translates into "Hey, shit head." People from various backgrounds may also misinterpret some logos without necessarily leading to disaster. Yet this highlights the importance of securing proper marketing feasibility analyses of projects.

CHAPTER 1, CLASS EXERCISE #4:

Ask trainees, or students, to search the Internet for logos they think could be misinterpreted.

PROPOSED QUESTIONS FOR DEBATE:
1. Are there good and bad logos? (Think of the Swastika.)
2. Do projects actually convey a specific image? (Think of the Eiffel Tower when it was first undertaken.)

1.12 What Are the Basic Marketing Models?

Marketing science rests on a certain number of models, each quite relatively simple. One of those models is what marketing professionals call the **4Ps of marketing** (product, price, place, and promotion). However, this is only one of many models. Some experts have ignored many of them in the marketing literature, although these models regularly govern marketing actions in actual markets. We review here some of the three main categories of models: operational, consumer behavior, and strategic.[32]

1.12.1 Operational Models

Operational models[33] differ from consumer behavior and strategic models in the sense that they describe the actions marketing experts (or market agents) take or intend to take.

1.12.1.1 Operational Models Pertaining to Marketing Experts

Typical operational models belonging to the marketing experts include the following (Table 1.8).

---

[32] Seldom do marketing books address these fundamental models. We are obliged to list and describe them because they are instrumental in the marketing feasibility of projects.

[33] We represent operational models by parallelograms.

Table 1.8   Four Marketing Operational Models the Marketers Use

This approach represents the core model of this book. Before preparing a marketing analysis, experts want to know if it is worth their while. The exercises we give during our seminars prepare readers to take this upstream thinking step, which few books recommend although it is essential.

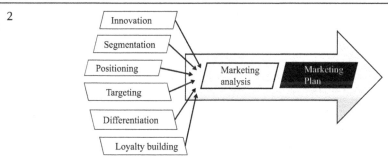

This constitutes the core of marketing analysis; we will define these six tools (innovation, segmentation, positioning, targeting, differentiation, and loyalty building) further along in Chapter 2. Generally, marketers must not complete a marketing plan (or strategy) without these six components.

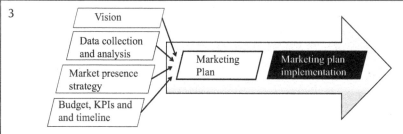

This is a simplified marketing plan. In essence, it must contain a vision extracted from the previous step named marketing analysis. Experts must thoroughly check the findings and expectations, and carefully assess clients and competitors. The last step of the marketing plan is precisely in line with the so-called iron triangle of project management: establishing a timeline, devising a budget, and setting the norms of quality and **key performance indicators** (KPIs). All the various models go through retroactive processes, with ongoing re-evaluation, until their dynamics create a final output.

(Continued)

Table 1.8 (*Continued*) Four Marketing Operational Models the Marketers Use

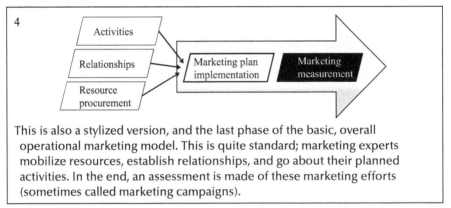

4

This is also a stylized version, and the last phase of the basic, overall operational marketing model. This is quite standard; marketing experts mobilize resources, establish relationships, and go about their planned activities. In the end, an assessment is made of these marketing efforts (sometimes called marketing campaigns).

Rightfully, some tutors or professors train their students in operationalizing Models 2, 3, and 4. However, Model 1 is the diving board upon which experts launch sound marketing efforts. Interestingly enough, marketing management science meets project management "science" in Model 3 (see Figure 1.5).

While each field of expertise—marketing or project management—uses its own tools and methods, it also has something in common with the other one… something at the heart of the marketing feasibility of projects: timeline, budget, and norms.

Coke, once again, provides a good example of the above operational models. Coke set to redefine itself with a new formula with the aim of increasing market share (Model 1). It went about estimating what could be in a new world with an improved Coke formula. Its marketing crew conducted a number of tests that confirmed people preferred the new formula to the old (Model 3), but the tests did not capture consumers' brand attachment. It then implemented the strategic plan (Model 4).

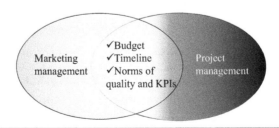

Figure 1.5 Areas where marketing and project management meet.

Note: Marketing and project managers clearly have a lot in common, and this is increasingly true. More and more companies hire marketing project managers under these, or similar, titles.

This perspective is very simple and summarizes basic operational models marketing experts use; about every other marketing effort falls into one or more of the above models.

CHAPTER 1, CLASS EXERCISE #5:

Prepare PowerPoint slides (at least one slide per element in the model), and fill them with real data found through simple research. This is especially useful for students with less knowledge or experience, and will be handy when comes time to do the feasibility exercises briefly presented at the end of this book.

1.12.1.2 Operational Models Pertaining to Consumers

The basic consumers' operating model is that of the purchasing process and its various expressions.[34] Many marketing books make sometimes the "arguable error" of calling this process the **consumers' purchasing process** but, in fact, at the point of purchase, the client is not yet a consumer. Only after having bought the product, and tried it, do purchasers become consumers through their own behaviors (e.g., experiencing satisfaction). Another common mistake is omitting the step of purchasing in the purchasing process, something that will baffle careful readers (this is an "arguable error"). The most useful model of the purchasing process, we find, is as follows (Figure 1.6).

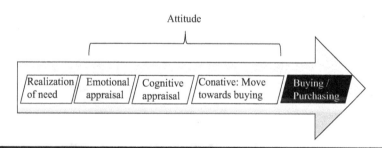

Figure 1.6   Industrial and regular consumers' purchasing process.

Note: Depending on the product, the purchasing context, and the personality of the buyer, emotional appraisal may be minimal and give way to more intense cognitive appraisal. Indeed, some models put the cognitive appraisal as preceding the emotional appraisal, especially in an industrial context; this makes sense.

---

[34] Such as the Fishbein model.

This model provides a succinct summary of the literature on the purchasing processes, as we like to use it in the area of project management.[35] Consumers first realize they have a need (or else develop a wish/desire/want). They then evaluate this need in terms of urgency, and seek information as to the means to fulfill it. They develop an emotional appraisal of such need and thus gather information. Of note, for products that have little share of the consumers' life, this step is often skipped; for example, buying a BIC pen is unlikely to generate emotions and the cognitive appraisal may be limited to ink color. Some products do not necessitate strong cognitive efforts; indeed, **impulse purchases** are merely a reflex to buy a product that engages few cognitive functions. Hence, the five-step model described here is flexible in its flow: Different products will engage each step at different levels of intensity. After the emotional appraisal is completed and cognitive appraisal performed, consumers make the decision to buy. This step is a behavioral one, and we call it the conative step. Together, we call the emotional, cognitive, and conative steps **attitude**. As can be seen, marketers can hardly change consumers' attitudes without addressing each of these three components. This is why it is so difficult to do; generally, marketers proceed one step at a time, such as by influencing consumers' purchasing patterns and hence their conative endeavors.

For project managers, the importance of such a model relates to the realization of the need: Indeed, project managers define projects in terms of a response to an opportunity. (Recall consumers' needs are the suppliers' opportunity.) Hence, this standard purchasing model highlights well a common point between marketing and project management: They both build on a market opportunity (see Figure 1.7).

Indeed, there is no marketing feasibility study that does not include a **needs' analysis** (or, more broadly, an opportunity analysis, for reasons we shall see in Chapter 4).

Figure 1.7 Another area where marketing and project management meet.

Note: In the end, a project revolves around consumers' needs. Project managers simply fulfill these needs by way of designing a project that will provide a deliverable.

---

[35] There are other marketing models, such as AIDA (attention, interest, decision, action), but we settle on the five-step model, which, in a shorter form (excluding the emotional appraisal), amounts closely to AIDA.

To turn the purchaser into a consumer or end user, another model is necessary: We outline it in Figure 1.8.

In four simple steps, this model captures what is happening in the market, and why we develop projects. Some forms of stocks are depleted[36] (e.g., source of energy, food, a bridge life span), triggering the purchasing process we have seen above. Purchasers become consumers, or end users, only once they use the products for what they are expected to do: Fulfill their needs. The process does not end there: There is a post-consumption experience that is inevitable, such as getting rid of packaging (garbage disposal) or filling out customer satisfaction surveys sent by the suppliers or sales force (e.g., Airbnb emails an appraisal form to its users after they have completed their stay). Only through this entire process can marketing and project managers fully grasp the dynamics of the market. Prospective clients turn into buyers when they decide to buy, and these buyers become consumers once they consume, and these consumers really only become clients once they process these steps at least a few times (see Figure 1.9).

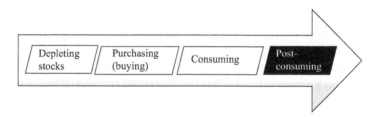

Figure 1.8   Need development in action.

Note: All needs are expressions of depleting stocks. When stocks (of energy, food, water, buildings, roads, etc.) diminish, survival commands their replenishment.

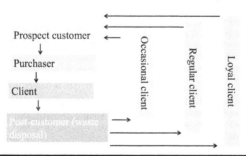

Figure 1.9   From prospect to loyal customer.

Note: The ideal customer is a loyal one. Loyal customers require less investment in promotion and will take part in promotion themselves by talking to their groups or references: family, friends, or club members.

---

[36] Experts sometimes refer to this phenomenon as the emergence of a gap between a current state and an end state, with the goal of achieving homeostasis (stability, at least temporarily) in the end state.

Within the framework of a project, this purchasing process only occurs once; repetitions infer that the project has effectively become an operation. (It is no longer a project as defined in Chapter 3.) Each step contains its own psychological components; typically, for example, depleting stock creates a certain level of anxiety that pushes individuals into action. The purchasing activity usually involves assessing the perceived risk associated with the product; what if the product does not deliver, or is faulty?

As can be seen, both marketing and project managers deal with something that is dynamic. In project feasibility analyses, as we shall see, anticipating the post-consumption step is one way of measuring potential positive and negative impacts, and may decide whether a project is a

- go ($\rightarrow$),
- a conditional go ($\leftarrow$), or
- a no-go ($\downarrow$).

Needs come in two forms: **real** or **latent** (**hidden**). Real needs are existing needs currently fulfilled by products, services, messages, or experiences supplied by the market. Latent (or hidden) needs are those that are yet unknown, dormant, unexploited, or in the process of being consciously apprehended. Project managers conceive projects, by definition, to respond to latent needs.

When an inventor discovers a latent need that is widely shared among the population, and thereafter develops a response, the potential for profit is considerable, in part because at that point there is no competition. Most modern, fast-emerging businesses have risen because a smart and creative individual has identified a latent need and found a way to satiate it. Facebook responded to the need to communicate as "friends" (or a need to socialize) in a society where people were becoming increasingly distant from one another, in part due to urbanization and job specialization (silo thinking). Amazon developed because of the need for people to receive desired products at home, given their lack of time to go shopping and the thrill with which purchases became much easier through the Internet. Such needs change with time, of course, and with societies. It is very telling that, in some parts of Africa, people have portable phones, yet not enough food.

1.12.1.3 Operational Models Pertaining to the Sellers

In our opinion, there are no clear models pertaining to sellers that have relevance to project management. The communication, purchasing, and psychodynamic models (covered in this book) suffice to infer the role of sellers as the latter are necessarily included as part of interaction with buyers.

1.12.2 Consumer Behavior Models

Consumer behavior models address psychological constructs rather than flow of activities.[37] Generally, marketers present models that include personal, promotional, and environmental stimuli, playing into consumers' psychology and characteristics, and buying decisions and processes. A few models emerge as part of the common knowledge base between marketing and project management: needs, psychological profiles, and **perceived risk**.

1.12.2.1 Needs

Many models discuss needs, including

1. Abraham Maslow's 1943 Hierarchy of Needs (physiological, safety, belonging, self-esteem, and self-actualization). This is the one we use in project management, in conjunction with economic theory.
2. Douglas McGregor's 1950 Theory X and Theory Y (geared mostly toward job satisfaction), and
3. Frederick Herzberg's 1959 two-factor theory of hygiene and motivational needs,[38]
4. Victor Vroom's 1964 Expectancy Theory (expectations leading to performances leading to outcomes),
5. David McClelland's 1998 Three Needs Theory,[39]

We can render needs using a mix of operational and conceptual modeling approaches, as shown in Figure 1.10.

Many books on project management make the mistake of stating that project management is about satisfying customers ("arguable error"). This is not true at all (and we shall see why in Chapter 3, which is on project management). The marketers' goal is to satisfy the longed-for customers. The hope is this will generate repeat business, encourage loyalty, and procure healthy benefits. In short, marketers identify an opportunity, develop their offer, coordinate their strategic activities, and satisfy customers while verifying they meet their quantitative objectives, such as profit and market shares. Project managers are only concerned with three things: timeline, budget, and norms of quality. Of course, meeting these should satisfy investors and please end users, but this is truly a by-product of the project's activities.

---

[37] We represent action steps (activities) by using parallelograms; psychological constructs use bubbles.

[38] Herzberg, F., Mausner, B., and Snyderman, B.B. (1959). *The Motivation to Work*. New York: John Wiley.

[39] McClelland, D. (1988) *Human Motivation*. UK: Cambridge University Press.

Figure 1.10   Needs and satisfaction.

Note: Marketing and project managers exert effort to influence consumers' satisfaction, by attempting to fulfill their needs. This attempt can go both ways: successfully or unsuccessfully.

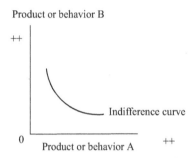

Figure 1.11   Needs seen under the eye of utility, by way of an indifference (utility) curve.

Note: We represent the trade-off between Products A and B, or Behaviors A and B, by a portion of a rectangular hyperbola and in line with our stylized approach.

One way to look at needs is to position them with respect to their fulfillment's utility, which is a core concept in economics. To understand how this works, let us put two near-substitute products[40] along a Cartesian map, as shown in Figure 1.11.

This graph shows the following: Consumers may elect to buy more of Product A and less of Product B, equally as needed, or less of Product A and more of Product B. Whatever their choice, they will reach the same level of utility: Both products have an equal capacity to fulfill their needs. Hence, economists call the curve that unites all the points between Product A and Product B an **indifference curve**, although we should really call it a preference curve because it shows what various sets of preferences exist. In a stylized fashion, it has the shape of a portion of a rectangular hyperbola, which means that the surface underneath each point

---

[40] If they were pure substitutes, the curve would in fact be just two lines joined by a 90-degree angle.

is equal whatever point along this curve the consumers sit on. Note this scenario applies to behaviors as well. Consumers may elect more of Behavior A (e.g., tax evasion) and less of Behavior B (stealing products off shelves), and vice versa. In either case, they will experience the same utility in the sense that will have saved (although illegally) more money, which was their initial goal. Indifference curves are not static. In fact, people become more and more sophisticated in their choices as time goes on and indifference curves tend to move away from their point of origin. For example, a teenage boy will be satisfied with a used car, but once he gets married at the tender age of 23, he will probably long for a more spacious, modern car. By 40, he would be proud to buy a fancy luxury car. Indeed, his basic need for transportation has evolved and become more sophisticated. This is true of about all products and all societies. Products A and B develop new generations, with increased features and better quality—just like iPad moved from generation one to two.

Also, recall the project management concept of utility drawback seen earlier (a utility drawback occurs when time is wasted, and/or extra costs are incurred, and/or quality is compromised). Figuratively, this utility drawback means that an indifference curve is getting closer to the point of origin, when in fact it should move away from it, thus reflecting improvements in living or project conditions.

There is more. While consumers may need a car, they may also need shelter, that is, a house. Therefore, a second indifference curve may sit on top of the first.[41] One way to rank indifference curves is to resort to Maslow's theory of the "ladder" of needs,[42] and assume that a rung represents each need. This way, the needs are various and become more sophisticated. Table 1.9 displays how this works.

This model is quite important in marketing feasibility studies; remember, however, that it assumes the market is risk-free. The more a project aims at satisfying a slew of needs, the more complex (and costly) it tends to be. The reverse is true: Complex projects are complex because they try to fulfill a wide variety of needs, and, in time, these needs tend to become more and more sophisticated. This is why a project such as building a space station is so complex: Project managers must ensure they meet all end users' needs, and this in precarious conditions.

Rule of Thumb: The more complex a project is, the more likely it contains points of vulnerability (POVs); hence, the higher the probability of failure.

---

[41] Utility curves have a certain number of characteristics; in particular, they cannot cross and more are always better (higher utility is preferred to lower utility).

[42] As seen previously, needs are ranked as physiological, safety, belonging, self-esteem, and self-actualization.

Table 1.9   Hierarchy of Needs and Indifference Curves and Their Utility

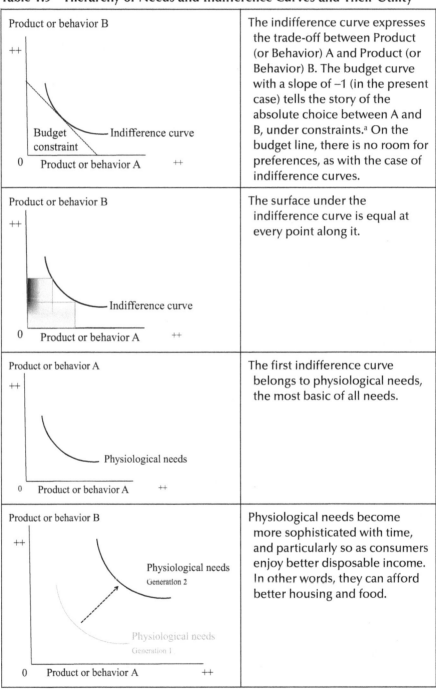

Product or behavior B … Budget constraint — Indifference curve	The indifference curve expresses the trade-off between Product (or Behavior) A and Product (or Behavior) B. The budget curve with a slope of −1 (in the present case) tells the story of the absolute choice between A and B, under constraints.[a] On the budget line, there is no room for preferences, as with the case of indifference curves.
Product or behavior B … Indifference curve	The surface under the indifference curve is equal at every point along it.
Product or behavior A … Physiological needs	The first indifference curve belongs to physiological needs, the most basic of all needs.
Product or behavior B … Physiological needs Generation 2 … Physiological needs Generation 1	Physiological needs become more sophisticated with time, and particularly so as consumers enjoy better disposable income. In other words, they can afford better housing and food.

(Continued)

Table 1.9 (*Continued*) Hierarchy of Needs and Indifference Curves and Their Utility

Product B ++ Safety Physiological needs 0 Product A ++	In fact, there are usually a number of concurrent indifference curves, because consumers must fulfill many needs at any point in time.
Product B ++ Belonging Safety Physiological needs 0 Product A ++	Indeed, we can express more needs by adding more indifference curves.
Product B ++ Esteem Belonging Safety Physiological needs 0 Product A ++	Here, all of Maslow's assumed levels of needs are expressed by way of indifference curves that move away from the point of origin.

Note: As can be seen from this table, four factors influence demand: (1) the presence of substitutes, (2) the level of utility of the particular need, (3) the budget constraints, and (4) the need being fulfilled (e.g., self-actualization).

[a] Any economics book will discuss such curve.

In general, we separate physiological needs into eight sectors of activity, for which databases are easy to access on about any government website.[43] These eight sectors are as follows:

1. Clothing;
2. Communication (e.g., portable phones);

[43] Example: www.bls.gov/news.release/cesan.nr0.htm. Accessed January 25, 2019.

3. Education;
4. Food, alcohol, or tobacco;
5. Health and beauty aids (HABA);
6. Shelter (house and house maintenance);
7. Sports and entertainment; and
8. Transportation.

CHAPTER 1, CLASS EXERCISE #6:

Fill in the following table: In rows are the eight sectors of activity, and in columns are (in that order) examples of a brand name or company, a core product emblematic of that brand. List three important attributes that come to mind for said product, then give the slogan that goes with the brand or product, and then, finally, give a list of major competitors. With respect to competitors, attempt to classify them according to their market presence (leaders, challengers, followers, or so-called nichers[44]).

This exercise is useful because participants can recognize major brands and products that occupy their mental space[45] on a daily basis, and can get a sense as to what a good slogan is[46] because it exposes them to the product's promise (or value proposition of a product, which is derived from the three main attributes, as we shall see in this chapter).

PROPOSED QUESTIONS FOR DEBATE:

1. How do you see your needs evolving? (e.g., for shelter through house purchases, for self-actualization through work)
2. Can an individual survive without satisfying the upper-ranked needs?

The above two need-based models only partly suffice to establish whether a project is feasible or not. It is a good idea to remember them, because they intuitively drive decisions marketing and project managers take. A project that cannot bring satisfaction to the end users has failed; one that does not respond to utility is also doomed. Thus, doing a needs' analysis within the framework of the marketing feasibility of projects must include these two models: needs versus satisfaction and needs as expressed by their utility.

---

[44] A niche is a very specific, highly defined market segment.

[45] Marketers wish to occupy two other spaces: the emotional and the physical. Together, the three spaces refer, of course, to attitude; hence, marketers attempt to control or influence consumers' attitudes by occupying their mental, emotional, and physical space.

[46] Good slogans express, directly or indirectly, the core business of the product and must be easily memorable (e.g., Volkswagen's "Das Auto" is punchy, short, and active; and Nike's "Just do it" is easily memorizable).

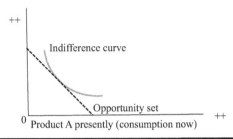

Product A in the future (consumption later)

Figure 1.12 **Opportunity set in a risk-free market (under certainty).**

Note: As this figure shows, three factors influence demand: (1) the urgency of the need represented by present and future consumption, (2) the capacity to fulfill the need represented by the indifference curve, and (3) the opportunity set.

1.12.2.2 Opportunity

Having developed a way to analyze needs, we can do something quite similar to understand the opportunity or **opportunity set**. A product is an opportunity to fulfill a need: If the opportunity is good, this means the need is satisfied in quality, quantity, or both. Thus, an opportunity set tells the story of needs fulfillment. Instead of comparing two products (or two behaviors) to which we adjunct a budget line, we take a single product (or behavior) and compare it in its present and future states. We are still here in a risk-free market (i.e., we operate under certainty[47]) as the consumers know how much interest they can earn on their invested savings (see Figure 1.12).

Consumers can spend their money now to consume Product A, or else they can put their money in the bank, earn interest, and consume Product A in the same quantity later, but with a gain in capital, that is, the interest earned on the money put aside at a known interest rate (also called a discount rate of, say, 5%). By consuming now, they gain an advantage: satisfying the need immediately before it becomes too pressing. By consuming later, they earn money, which they can spend to buy, say, more of Product A than initially needed.[48] Thus, consumers travel along this type of indifference curve by trading off the present and the future consumption. However, their financial capacities remain limited, and we represent this by the opportunity set: If the slope of this set is −1, then consumers must give up one unit of future consumption if they use one unit of

---

[47] "Under certainty" means that we know what the future state of the product will bring about. If the future is uncertain, then we cannot assert that future consumption will actually occur, or occur the way we anticipated. In projects, the future is always uncertain; thus, the present model is a simplification of the reality.

[48] In economics, we refer to the marginal rate of substitution (MRS), which changes along the indifference curve.

present consumption. Otherwise, they must give up one unit of present consumption if they delay consumption.

The important point here is that we have a way of identifying the opportunity and, if you recall, as we saw in the previous subsection, we already have a way of modeling needs. Consumers are concerned with their needs; suppliers (marketers and project managers) reposition these needs in terms of present and future states and develop an opportunity set accordingly. Both the needs as expressed by Product (or Behavior) A *versus* Product (or Behavior) B, and by present *versus* future states, illustrate, in fact, **consumers' total preferences**. Consumers prefer Product A to Product B (or Project A to Project B), and they prefer it now rather than later. For example, a group of hungry consumers will prefer Product (or Behavior) A to Product (or behavior) B—like eating apples *versus* oranges—and will prefer to consume them now rather than later. A government faced with two infrastructure projects will favor Project A *versus* Project B given its present budget and electoral pressures, and will prefer completion of Project A sooner rather than later.

In the context of a realized project, we therefore posit that consumers base their decisions to consume a product on at least seven criteria:

1. The availability of substitutes,
2. The type of need (as seen along Maslow's hierarchy of needs),
3. The level of utility,
4. The budget constraints (or the capacity to buy the coveted product),
5. The urgency of the need (or how pressing it is, represented by present *versus* future consumption),
6. How well the product quantitatively and qualitatively responds to the need, and
7. The opportunity to buy and consume said product (the opportunity set).

This is equivalent to saying that, ultimately, consumers demand a product (or a project) based on at least these seven criteria. Mathematically, we could express this as follows (Equation 1.1):

■ Equation 1.1: Decision to buy and consume

$$DC(\text{decision to buy and consume})$$

$$= f\left(\beta_1\text{Substitutes}, \beta_2\text{Type}, \beta_3\text{Utility}, \beta_4\text{Budget}, \beta_5\text{Urgency},\right.$$

$$\left.\beta_6\text{Response}, \beta_7\text{Opportunity}\right)$$

Knowing this stylized function and the two types of curves (the near-substitute products A–B and present-future indifference curves, as well as the budget and opportunity set curves) provides plenty of information to both marketing and project managers. In fact, project managers deal with products that they have not yet realized; hence, it is a projected product (in the future, of course), a deliverable. This does not prevent them from considering other current products that may cast a shadow of doubt over the success of the future product the project aims to materialize. It does not prevent them either from evaluating how pressing the completion of the project is.

1.12.2.3 Supply and Demand Curves

The above discussion entails someone who is out there willing to offer Product A and/or B, or to encourage Behavior A and/or B. That someone, of course, is the seller (supplier), which, in our discussion, amounts to marketing and project managers. As mentioned before, marketing borrows heavily from economic science, as is the case again here. Traditionally, economic science positions the quantity of goods demanded on the X-axis instead of the Y-axis, where it truly belongs as consumers choose a product based on price and not the reverse (Table 1.10).

As discussed, neoclassical economists assume that consumers are rational and that they have all the necessary information to make the best decisions in order to fulfill their needs. In addition, they take for granted that consumers accept the price tagged on the product, that is, the price that appears on the sticker affixed to the packaging. In reality, however, consumers go through a mental exercise by which they take into account the cost of pursuing additional information, including among competitors, and the cost of actually going to the point of sale. Even a virtual point of sale, such as an online store, implies a cost: Customers have to pay a monthly fee to maintain their Internet connection. Hence, the buyers calculate the expected **benefits** they think they can get from the product and the **sacrifices** they incurred in acquiring it,[49] such as Internet subscription, the cost of the travel to the point of sale, and the time wasted searching for information, comparing stores, and so forth. Customers will normally only buy a product if the benefits outweigh the sacrifices. Hence, in Table 1.10, the requested quantity is indeed a function of price, but not only of the stickered price; it is a function of the overall price. We formulate this in a stylized manner as follows:

- Equation 1.2: Demand (short version)

$$\text{Demand} = f\left(\text{Quantity} \,|\, \text{Price}\right) = f\left(\text{Quantity} \,|\, [\text{benefits} - \text{sacrifices}]\right)$$

This reads as follows: Customers demand quantities of products according to (a given) price. Put differently, the fewer sacrifices they have to incur and the more benefits they stand to gain, the more likely they are to buy in large quantities.

---

[49] Some books refer to cost, but this reduces the way customers experience the buying process as they don't all put a monetary value on their efforts; in fact, in some cases, this may be impossible. Rather, customers estimate what they give up (sacrifice) in order to obtain what they wish for, such as time, money, and relationships.

Table 1.10 Supply and Demand Curves in Various Contexts

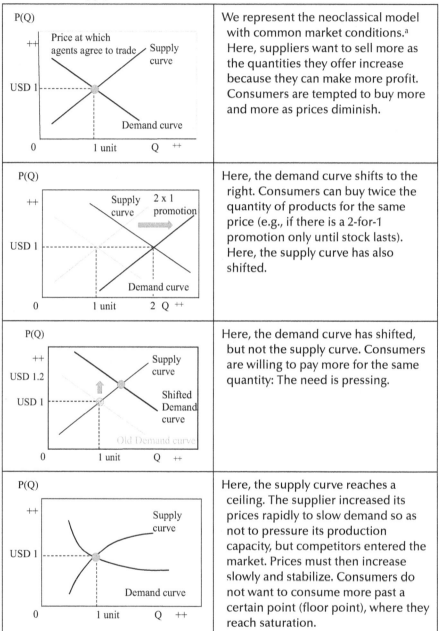

	We represent the neoclassical model with common market conditions.[a] Here, suppliers want to sell more as the quantities they offer increase because they can make more profit. Consumers are tempted to buy more and more as prices diminish.
	Here, the demand curve shifts to the right. Consumers can buy twice the quantity of products for the same price (e.g., if there is a 2-for-1 promotion only until stock lasts). Here, the supply curve has also shifted.
	Here, the demand curve has shifted, but not the supply curve. Consumers are willing to pay more for the same quantity: The need is pressing.
	Here, the supply curve reaches a ceiling. The supplier increased its prices rapidly to slow demand so as not to pressure its production capacity, but competitors entered the market. Prices must then increase slowly and stabilize. Consumers do not want to consume more past a certain point (floor point), where they reach saturation.

[a] Technically, the graph reads "If I sell X amount of units (Q), I will obtain this price [P(Q)]."

Economists also assume that the market is perfectly competitive; there are always suppliers ready to provide whatever consumers need and demand. Finally, and as previously discussed, they generally assume consumers are not time-sensitive; they posit they will wait in line for as long as it takes to buy their basket of goods without getting frustrated. In a sense, this contradicts the notion of opportunity set, which considers time as a key factor and the fact that customers do get frustrated and tired. (Think of waiting in line for hours to buy your groceries!) In general, customers get frustrated regarding the time it takes to access the needed products and the quality of the goods they buy.[50] This is something we will see again when we discuss projects.

CHAPTER 1, CLASS EXERCISE #7:

Discuss the different possible scenarios with respect to supply and demand curves. Provide market examples.

PROPOSED QUESTIONS FOR DEBATE:
1. Should customers be allowed to ask for a compensation when they wait too long?
2. Are there needs for which you cannot wait forever?
3. What are the underlying assumptions of a given project? The more one costs, the more similar ones will be built? However, the less a project costs, the more consumers will want to see it realized?
4. What main factors drive the demand for a given project? Utility, price, opportunity?

Of course, assuming eternal patience is not realistic, but such simplification is nevertheless useful to express market dynamics.

When put in different combinations, the above-mentioned seven criteria (substitutes, type, utility, budget, urgency, response, and opportunity; see Equation 1.1) move the demand curve to the right, to the left, up, or down. For example, a very pressing need, in which time is of the essence and for which the opportunity set touches the indifference curve closer to present consumption, will make the price nearly inconsequential. Consumers facing a dire, threatening health condition will pay a very high price for medication that could be bought at a much cheaper price should there be no urgency.

The supply and demand curves are generally presented as straight lines but, in fact, they can (and are often) be curved. In exceptional cases and depending on

---

[50] Around the world, increases in the cost of living regularly drive people into the streets to protest against their inability to make ends meet at the end of each month.

the same set of factors, we can reverse the slopes of the supply and demand curves. For example, in the case of exceptional (rare) products sold at an auction, the more expensive the product is (e.g., a Monet or Van Gogh painting), the more determined buyers push prices up. The same phenomenon occurs in certain real estate markets. Note that the market is near, but not perfect: Substitutes are at play here, as predicted by Equation 1.1. The rarity of the product means there are few or no substitutes at all. Time is also of the essence, of course, because auctions operate under intense time pressures. This means the slope of the supply and demand curves may theoretically vary from zero to infinity in either direction (positive or negative).

In addition, experts usually use supply and demand curves without maximum borders. However, this does not actually reflect reality in the context of projects, as they are bound by time, budget, and norms of quality. Hence, for the sake of modeling supply and demand in that context, we always insert the supply and demand curves inside Edgeworth boxes (as we did in Table 1.10), which sets limits on their evolution. For the same reason, we will from now on examine utility (indifference) curves for both pairs of products or behaviors (A and B) as well as products in two states (present and future) within Edgeworth boxes.

The notion of quantity seldom applies to a project. A project is unique; thus, one cannot offer quantities of the same project. For a given quality, drivers of a project are usually its costs (budget) and timeline to realization (present *versus* future consumption).

1.12.2.4 Profiles

We have just seen that preferences are important. Preferences vary with cultures or, more precisely, within groups of consumers or between individuals. Thus, another crucial element of marketing theory with respect to consumer behaviors is profiles.[51] All marketers do profiling one way or the other, if only because they must understand their customers and tailor their products accordingly. There is a fair number of contrasting perspectives on personalities, including the so-called Big Five,[52] the various ones described by *Diagnosis and Statistical Manual of Mental Disorders-V* (DSM-V),[53] and many others. These do not offer the advantage of capturing personalities in a way useful for the project management environment. According to the attachment theory,[54] which is a step closer to being useful in marketing and project management, there are four main profiles: anxious, avoidant, hostile, and secure (stable).

---

[51] One theory often used in marketing books is that of AIO, for activity, interests, and opinions.

[52] Goldberg (1990). According to the Big Five personality traits, a person can be agreeable, conscientious (meticulous), extraverted, opened to experience, and a neurotic (nervous) geared toward sociability and work skills overall.

[53] The *DSM-V* (2018) deals with disorders (e.g., antisocial, borderline, histrionic, narcissistic).

[54] Bowlby (1973).

For the sake of our marketing perspective, we believe it is better to focus on the four basic **coping mechanisms** we know exist in nature and associate them with a personality type[55] (see Figure 1.13).

Anxious persons doubt themselves and their surroundings. They exhibit blends of positive and negative emotions, are submerged by internal and external conflicts, and exhibit fear of abandonment, disapproval, humiliation, unfairness, or rejection.[56] They may show gratitude or forgiveness, but soon resort to negative self-referential feelings. Anxious clients pose a lot of questions, often in a disorganized manner, and some do not seem to listen to the answers. Their understanding of their own needs is blurred by their anxiety, which tends to make their fulfillment urgent no matter what.

Avoiding types, as the name indicates, are poor at socialization and do not deal with problems at hand. They are socially disengaged, become easily bored or tense, and ignore the feelings of others. Avoiding clients walk into stores only to try to get out; they do not help the salesperson who genuinely tries to assist them. They are evasive when answering questions. They build some frustration from the fact that they are not capable of finding ways of fulfilling their needs.

We divide hostile personalities into two groups: defensive (controlled by the medial hypothalamus of the brain) and instrumental (controlled by the lateral hypothalamus).

Defensive personalities react rather than act. It is a bit like a cat who scratches at a foot that has inadvertently trod on its tail. Defensive customers think the salespeople are out to get them, to deceive them, and to abuse them. They are not proactive in their search for the very products and services that could fulfill their needs.

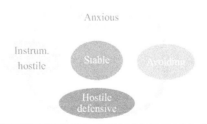

Figure 1.13   Consumers' personality types.

Note: In our model, based on the works of Bowlby (1973), among others, we see four personality types according to the four basic coping mechanisms humans resort to. When harmonized, these four coping mechanisms create a "stable person."

---

[55] It is out of the scope of this book to go at length into personalities and their underlying neurobiological networks.

[56] Known as the five core emotional childhood wounds.

Instrumental hostility is a delayed response; that is, it is delayed defensive hostility, which involves the planning of future retaliation. Hostile people of this type may develop a hatred for a particular product, company, or brand name; sabotaging, bad mouthing, and like behaviors are part of their portfolio of tactics for exerting revenge.

When these four personality mechanisms work in tandem, we consider individuals as being in a stable state.[57] As such, stable personalities do not exist *per se*; rather, individuals show stability when they know to resort to the appropriate coping mechanism depending on the circumstances they face. Some circumstances call for defensive reactions (e.g., one has to push off someone inadvertently walking on their toes), instrumental hostility (e.g., delaying a response because the immediate threat is too overwhelming), avoiding (e.g., turning away from a speeding car when crossing the street), or anxiety (e.g., in taking care of young children who venture in hazardous terrains).

CHAPTER 1, CLASS EXERCISE #8:

Position yourself along the various personality types/coping mechanisms. You can keep this information for yourself, or share it.

PROPOSED QUESTIONS FOR DEBATE:

1. Are there any marketing situations in which you have exhibited hostility? (e.g., when facing an impolite salesperson or when waiting in too long a line)
2. Should there be laws to punish unfriendly consumers?

Marketing and project managers long for clients and business partners who are stable individuals, as experts have found them to share a number of healthy characteristics. They are focused and know what they want. They are well balanced intellectually, emotionally, and behaviorally; they trust and cooperate. They keep their relationships longer than avoidant or anxious individuals. They display little hostility. They have a functional level of self-confidence and don't long for conflict (see Chapter 4 for a discussion on conflict). They have a strong sense of attachment toward their goals and their surroundings, and display a positive vision of the world.

Furthermore, stable individuals excel in interpersonal relationships. They are flexible and respond to the needs of others. They focus first on people rather than on tasks; they are predictable and use heuristic logic, taking into account a large view of the world rather than a closed, negative one. They do not tend to display

---

[57] Mikulincer and Shaver (2007).

episodic mood changes or unexpected variations in their decisions. They are usually better at developing their own talents and tend to excel academically and/or in extracurricular activities. They are able to concentrate, are not bothered by negative thoughts or souvenirs, and do not feel victims of events or of others.[58] Rather, they show their cognitive, emotional, and conative prowess. Table 1.11 reports on the difference between stable individuals and others.[59]

Marketers and project managers try to identify stable customers/end users because they are the most rewarding in the long term. Yet, marketers also bet on the frenzy of people, at least in the short term, or on a much-focused basis. Black Friday sales have the reputation of driving people completely insane, even inciting violence to the point that some deaths have occurred during frantic shopping sprees. Similarly, marketing experts built fancy, sweet deals during the predatory mortgage wave in the years leading to the Global Financial Crisis of 2007–2009, which created millions of victims in the United States and around the world.

Table 1.11   Traits of Stable People

Behavioral Trait	Valence Evaluation	Associated Main Psychological Construct
Trustworthy	0.95	Trust
Caring	0.91	
Responsible	0.91	
Confident	0.63	
Attractive	0.79	
Emotionally stable	0.91	Equilibrium (balanced)
Sociable	0.90	Cooperation
Intelligent	0.70	
Unhappy	−0.70	Hostility
Aggressive	−0.75	
Mean	−0.78	
Dominant	−0.30	
Threatening	−0.78	
Weird	−0.85	

---

[58] Magdol et al. (1998).
[59] Todorov and Engell (2008).

Many books dedicated to consumer behaviors skip the discussion on personality; yet, as we shall see in the next section, personalities affect the overall buying experience.

1.12.2.5 Apprehension and Perceived Risk (Threat)

One of the core drivers of personality and key concepts in marketing, finance, and psychology is **perceived risk** (or perceived threat). In nature, all animals constantly monitor their environments and debate between foraging (searching to fulfill their needs) and the possible presence of predators.

All products have an inherent potential for failure: They may not do what they promise to do, they may break sooner rather than later, or they may actually be harmful (e.g., lead in blinds, nicotine and harmful chemicals in cigarettes). Customers are thus completely justified to be worried of potential dangers associated with the products they buy. A high-perceived threat means that stakeholders experience little trust. Investors who do not trust their financial advisers are unlikely to cooperate or invest more money in the market through their services. If the customers think sellers are unfair (e.g., when offering them a price that is higher than what recent TV ads promised), they will likely disengage. However, should these customers be rewarded with healthy returns on investments effected through their financial adviser, they will most likely perceive less threat in future investment projects.[60] We illustrate this dynamic quite simply in Figure 1.14 (see Appendix 2). Obviously, stable personalities favor such a positive cycle, whereas customers who are excessive in their levels of anxiety, tendency to avoid contact, and defensive or hostile behaviors are more likely to stick to a vicious circle.[61] The more displeased they are, the less friendly they become.

We read Figure 1.14 as follows: Consumers sense a need, which calls for action. They look for the products that may fulfill these needs, but if there are more sacrifices than benefits associated with obtaining the product they long for, they actually perceive it as a threat. When consumers perceive risk with a product or feel threatened by a salesperson, for example, they become less inclined to buy the coveted product. Their level of trust diminishes.[62] If, however, the perceived risk is low, then there are no reasons to doubt or be suspicious. Trust increases, so market agents have no reason not to cooperate—quite the opposite. There is a positive link

---

[60] The correlation between perceived risk and trust is not linear and probably resembles the financial model of risks (expressed in a percentage) *versus* number of stocks held where these are indicative of the trust investors have in the market. This generally forms the shape of a rectangular hyperbola (Elton et al., 2011, p. 60). The other correlations between various variables of the model have been found in numerous studies (including ours) to be linear.

[61] The worst customers are the instrumentally hostile ones.

[62] Trust is formed by four components: affinities between sellers and buyers, benevolence, abilities (competences), and integrity.

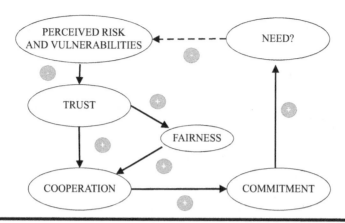

Figure 1.14   The apprehension model.

Note: Figure 1.14 has the advantage of hinting at the kind of personality involved in the process expressed by the model. For example, an anxious person would certainly have high levels of perceived risk and vulnerabilities and may display little trust toward others.

between trust and cooperation. If buyers feel they are being treated fairly, then this adds to the motivation to cooperate. Hence, there are positive links between trust and fairness and between fairness and cooperation. Cooperation[63] is more likely to reap the expected rewards than if the market agents fight and refuse to work together. There is a positive bond between cooperation and reward. As the wallet gains in weight due to the accumulation of profits and benefits, there are no reasons to doubt the other agent (the salesperson) further, and thus perceived threat (or perceived risk) diminishes. There is a negative relationship between reward and perceived risk; when one goes up, the other goes down. Again, if perceived risk, for whatever reason, starts going up, trust inevitably diminishes. There is a negative (non-linear) relationship between perceived risk and trust.[64]

The model in Figure 1.14 is somewhat akin to other purchasing models to which marketers resort, such as the AIDA model or, for a more complex one, the AICIEP (awareness, interest, consideration, intent, evaluation, and purchasing), which is useful in industrial purchasing.

This model expresses the core mechanism to which sellers and buyers resort, and neurobiological evidences tend to support it. Obviously, stable individuals efficiently measure how much perceived threat should be reasonably expected (defensive hostility), how much anxiety (trust) to feel, how distant they should

---

[63] We express cooperation with four variables: flexibility, exchange of information, joint problem-resolution, and orientation (acting for the benefit of the other agent).

[64] The triangle of trust-fairness-cooperation forms the behavioral core and, equivalently, describes the atmosphere of the relationship. Reward is external to this triangle.

be from the trade taking place (avoidant), and how to plan their future moves (instrumental hostility).

Put differently, defensively hostile customers will be overly suspicious and move away from buying the products they nevertheless need. Anxious people will tend to distrust and become ambivalent in their choices. Avoidant customers will refuse to engage in a relationship, even a short-term one, with the salesperson. Instrumentally hostile customers will seek revenge because they feel they have been somehow mistreated. Experts deem those customers who exert a correct balance between these valuable coping mechanisms are stable as they foster a virtuous circle as drawn in Figure 1.14.

All marketers work with this model in mind, consciously or not, as it has the advantage of being simple and easy to use when developing a marketing plan. In short, the marketers, as well as project managers for that matter, want the customers/end users to have a positive, nonthreatening image of their products or projects. They want them to develop some form of attachment to their company and to their brand name. They certainly want to convince consumers they are treated fairly (a sense of unfairness inevitably leads to anger). They long for cooperative customers, and do what is necessary to be friendly and make them feel at ease. They reward them with products or projects that offer high utility, something that preferably exceeds their expectations of needs' fulfillment. The more they achieve this, the more consumers become loyal.

Figure 1.14 also provides a simple rendition of the key mental structure of consumers, or, put differently, of the mechanism of consumer behaviors. Consumer behavior theorists have derived a number of behaviors that relate to Figure 1.14; for example, **cognitive dissonance** refers to the sentiment buyers have when they arrive home with the product they just bought and feel uncertain about their purchase after all. They may even consider returning it without knowing exactly why. Somehow, this dissonance creates an imbalance, and balance is part of the above model. Marketers can also explain other behaviors by resorting to Figure 1.14. Over-consumption talks of erroneous assessments of the needs (a flawed "?" in the drawing). Addiction evokes high dependency on products that is detrimental to the individual's health. Abusive behaviors, such as theft, excessive anger, sabotage, and the like, imply a sense of frustration, which means that customers have not been fulfilled their needs and that they feel they have paid for nothing. They feel they have been tricked, or treated unfairly. Boycotting and other such behaviors also talk of dissatisfaction, but without the anger component. Taboos are filters that affect needs' assessments, just as are cognitive or emotional biases.

Project managers take for granted that consumers would accept their projects. The community where the managers "operationalize" them do not usually deal with such behaviors, but circumstances may change. A pressure group formed on disgruntled potential end users may all of a sudden pop up and try to interfere with the project. Hence, understanding how stakeholders behave is always a good asset in project management.

1.12.3 Strategic Models Used by Marketers

Armed with the above-described models, marketers conceive and implement marketing plans in order to boost sales and profits. To achieve this, a limited number of strategic models suffice, which we describe below.

1.12.3.1 Profit-Seeking

We decline the core models linked to profit-seeking (for profit-seeking firms) in four forms, as shown in Figure 1.15.

We can link these models one to the other, but it is much simpler to separate them as we do below, as this is usually how business managers think. The first model states that as marketers develop products that fulfill the needs of their customers (thus, these products present the expected utility), these customers are likely to be satisfied. It is not a guarantee, but it is a fair bet; some customers will never be satisfied, no matter what, especially the "non-stable" ones. It is also a safe bet to assume that satisfied customers will buy the same product again, even though some will seek products from competitive brands for the sake of diversifying their purchasing experience.

Most companies spend money to advertise their products in order to occupy the emotional, the mind, and the behavioral spaces of their customers. Yet, no study has ever convincingly proved, in some form of mathematical function, that advertising does indeed increase sales. Finally, business people theoretically invest profits to nurture the growth of their company.

Needs' fulfillment and customers' satisfaction certainly belong to the R&D and marketing departments. Customers' satisfaction and repeat purchases are of immediate concern to the marketing and sales departments, while promotion and profit generation interest the marketing and strategy departments. Finally, increased sales and profits encourage growth and are key concerns to top management and investors.

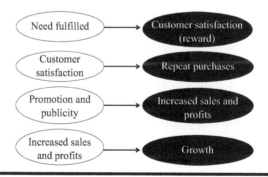

Figure 1.15   Four core models linked to profit-seeking.

Note: These four models are used in the daily life of about any business, presented here in their simplest form.

1.13 Conclusion

This chapter presented an overview of marketing and focused on the theoretical aspects with which project managers should acquaint themselves. We clarified some key concepts and expressed concerns about some preconceived ideas that have a dubious application in the field of project management. We presented the basic models that consistently drive marketing theory, an approach we believe is rarely found in marketing books. However, these models have the advantage of simplifying marketing theory and of showing how it relates to challenges faced by project managers.

It is clear from our discussion that marketing borrows heavily from various disciplines, including psychology, sociology, economics, and even the military. We trust Chapter 1 raises enough interest among readers less familiar with marketing to want to know more and seek further evidence of how marketing activities affect our everyday lives. The next chapter delves into marketing management, that is, how we apply marketing concepts as part of an overall business strategy.

1.14 Mind Teasers

Readers may use the mind teasers as questions in preparation for an examination or quiz.

1. Define
   a. attitude by its three components,
   b. marketing in the simplest format,
   c. perceived risk (or perceived threat),
   d. real and hidden (or latent) needs,
   e. the mechanism of gray markets, and
   f. the term "market."
2. Describe
   a. the five things we can put into the market,
   b. the four main eras of marketing,
   c. the four market agents, and
   d. the three main types of relationships between sellers and consumers.
3. Draw and briefly describe
   a. the non-industrial consumers' buying process,
   b. one of the four core models linked to profit-seeking,
   c. one of the four marketing operational models marketers use, and
   d. the apprehension model.
4. Explain
   a. the link between demand, quality, benefits, and sacrifices, and
   b. the process by which prospects become loyal customers.

5. Give examples of, and briefly describe,
   a. at least three personality profiles,
   b. at least five traits of stable clients,
   c. at least six criteria customers may consider when planning to buy a product,
   d. examples of behaviors expected from marketing experts,
   e. examples of organizations defending the interests and rights of consumers, and
   f. examples of reprehensible marketing practices.
6. List the characteristics of consumers' and industrial buyers' behaviors.
7. Plot
   a. an indifference curve (in a complete graph) and
   b. an opportunity set (in a complete graph).
8. True or False? The suppliers' motivation is the buyers' opportunity
9. What are/is
   a. the eight sectors of activity typically found in governments' databases,
   b. the five fundamental rights of consumers,
   c. an oligopoly, and
   d. a POC?

Chapter 2

What Is Marketing Management?

The archer represents the marketers, who use four arrows to target their audience: product, price, place, and promotion.

2.1 Introduction

We are now getting into an important aspect of marketing: marketing plans. Many people unfamiliar with marketing theory confuse such concepts as segmentation and positioning. For the sake of our effort at mixing marketing and project management, we classify the marketing plan into six categories: innovation, segmentation, positioning, targeting, differentiation, and finally, loyalty building. Marketers must set their goals before implementing the marketing plan, often by cooperating with project managers and investors. Marketing plans propose programs which themselves include activities (such as a celebration). Inevitably, marketers evaluate their efforts; that learning becomes part of the book of knowledge every project generates. Such efforts involve collecting primary data (data that does not already exist) and secondary data (recorded in one form or another: articles, reports, government websites, etc.). Collecting this data involves identifying the research questions (what it is we are trying to find out), setting the proper research method and ethical steps to collect the data, analyzing the data, and then offering or deciding upon actions steps, given the allocated budget. Elements of the target market to measure include the anticipated number of end users, likelihood of repeat usage, project/brand attachment (see the section on brands in this chapter), and the like. In many modern organizations, clients drive projects and innovations as opposed to old-fashioned management, whereby top decision-makers dictate their tastes to end users. As previously seen, marketers and project managers wish to make potential end users aware of the deliverables to increase interest, to position themselves within their consideration set, and to encourage adoption as much use as possible. The idea is always to boost **customer lifetime value** (**CLV**), that is, the amount of profit that can be gained from customers' lifetime usage of the deliverable. Managers sometimes devise various strategies to maximize use, including lending, offering different uses than the original (e.g., a sports venue turned into a museum), or rental or storage. In short, the marketing of the deliverable can be innovative!

The learning objectives of this chapter are to understand what marketing management is, to recognize marketing management in daily life, and to make the link between project management and the six core components of marketing management: innovation, segmentation, positioning, targeting, differentiation, and loyalty building.

2.2 Six Components of Marketing Management

We prefer to address the six components in the above-mentioned order but, in reality, there is many back-and-forths between them (see Figure 2.1).

Simple definition: **Marketing management** is about innovation, segmentation, positioning, targeting, differentiation, and loyalty building.

Figure 2.1 The six components of marketing plans for marketing feasibility of project analysis.

Note: The six components of this model encompass all there is to address when doing marketing feasibility of project analysis.

2.2.1 Innovation

Innovation is an intrinsic characteristic of projects. It necessitates walking off the normal paths set for doing things: This is what makes projects so exciting. When discussing research, some authors have said "… any significant research requires that one tries out new paths and faces ambiguity to define new variables …."[1] Indeed, projects are spoiled with ambiguity and unknown variables, which in turn make them good targets for points of vulnerability (POVs) to carve a niche, mature, and take their toll on managements' and team members' efforts.

Over the years, countless innovative products have improved and led way to more sophisticated versions. Examples include typewriters *versus* computers, blackboards and chalk *versus* whiteboards and marker, public telephones *versus* mobile phones, and hotels' metal keys *versus* electronic card keys.

> Simple definition: **Innovation** is the skillful art and science of proposing new ways of responding to latent or existing needs.

Patents and copyright certificates are indicators of the innovative aspects of a project. A truly innovative product must meet three criteria:

1. It must be unique,
2. It must be useful, and
3. It should not consist of existing scattered elements that the creators put together without adding real value (functionality and/or design).

---

[1] Parkhe (1993, p. 229).

Backward invention or engineering consists in decomposing an existing product or strategy to identify its components and their interactions to then emulate or improve the finished product. Japanese engineers used such process extensively in the 1970s in such sectors as automotive.

Research and development (R&D) and marketing departments often work hand in hand to adapt an existing product or produce a new one, as well as to adapt an existing promotional program or create a new one. This may apply to both the project and the project's deliverable.

Innovators base their innovations on an underlying concept. For example, ink, paper, parchment, pens, and even the use of sand or rocks all convey the concept of written communication and speak of the receivers' ability to read a given message. Similarly, airplanes, balloons, delta-planes, kites, and zeppelins all share flying as an underlying concept; or more precisely, the capacity to benefit from air flows to travel. A large number of innovations spur out of a source inspired by nature. The invention of Velcro, for example, owes to the plant *Arctium lappa L.*

A project is innovative by nature, but in the context of project, it is so only when it displays the following elements:

1. It has some kind of evolutionary goal,
2. It can deploy in various forms,
3. It has various levels of evolution,
4. It implies a compromise (in a project, the compromise is between time, costs, and norms of quality),
5. It has likely been inspired by a source such as nature,
6. It is a response to a problem, need, desire, or source of discomfort (e.g., social or technological), and
7. It has a certain utility (not all innovations are useful).

The typical breakdown of innovations is continuous (slight improvements to an existing product), incremental (considerable changes to an existing product), or disruptive/revolutionary. Of course, this applies to projects as well. Not all projects are revolutionary, even though those some catch both our attention and our imagination. More and more projects rely on digital platforms. Companies like Airbnb, Amazon, Facebook, and Uber have all reached outstanding market capitalization (no less in outstanding record time) by developing their offers on digital platforms. They all started as projects that quickly produced a deliverable that consumers needed, *en masse*.

Not all new products are successful, however. Causes of failure include erroneous assessment of the market, a poor launch, lack of support, and resource shortages. In part, this list emulates causes of project failure and highlights how important it is to determine the feasibility of a project. Managers must screen the innovative idea, develop the concept, proceed with product development and market evaluation, develop full marketing and business plans, and resort to pilot tests prior to full launch or investment in the project. A sound marketing plan will identify early adopters of the proposed innovative deliverable, and those who are likely to embrace it.

Product creators balance functionality (including compatibility) and design (including ease of use). From a marketing point of view, they seek to gain a competitive advantage they can effectively communicate. Products that succeed, such as many Apple products, excel at mixing both contingencies by way of innovative solutions. An ideal scenario occurs when these experts achieve 100% functionality while combining high-quality design with superior ergonomic and hedonic qualities. Under this scenario, we define quality as the sum of functionality and design (regardless of cost),[2] whereas value is quality over costs.

Costs exert a negative influence, yet achieving a superior trade-off between functionality and design usually requires much research (complexity comes from such conciliation). Solutions are not always easy to find. This proposition contains the critical elements potential investors worry about when promoters first present their project. It precedes the actual **project charter**, which once approved, "formally initiates the project."[3]

Thus, the **initial value proposition** of a project is the first formal, most succinct representation of the project's plan. In a sense, all plans start with the initial value proposition for a project. What value will the product (deliverable) generated by the project bring? The need for a prefeasibility study derives from initial value propositions (see Chapter 4).[4]

Experts have proposed steps to ensure consumers/end users adopt a new product. Customers must

1. Become aware of the product,
2. Develop an interest,
3. Seek more information,
4. Evaluate and sample the product (if needed) or prototype,
5. Buy said product as a gesture of early temporary adoption,
6. Repeat purchases once they are satisfied, and
7. Confirm the value of the product by becoming loyal.

Factors that facilitate product adoption are relative advantage ([benefits-sacrifices] gained compared to similar products), credibility (observable, positive, socially shared attributes), ease of trial, matching (e.g., with lifestyle), and simplicity.

---

[2] There are several known systems used to measure quality, such as Six Sigma and ISO 9000 norms.

[3] Project Management Book of Knowledge (PMBOK) 5, p. 67.

[4] There are many tools available to foster innovations such as conceptual maps, brainstorming, and the like.

CHAPTER 2, CLASS EXERCISE #1:

Imagine an innovative product, even one that doesn't make sense or that you've only dreamed of. Try to write an initial value proposition, enticing potential clients to buy your product.

PROPOSED QUESTIONS FOR DEBATE:
1. Is it acceptable to over-emphasize the benefits of a product?
2. Should innovative products not be protected by patents or authors' rights?

2.2.2 Segmentation

Many people think one can segment products. This is an "arguable error," at least from the point of view of this book.

Simple definition: **Segmentation** is about grouping people in an organized way.

Segmentation is about people and people only.[5] Segmenting is simply creating one or many groups that are as different from one another in order to better market products (or projects, of course). One can segment according to three formats. The first, **micro-segmentation**, focuses on a small percentage of customers (or potential customers, of course) compared to the overall population. When Michael Dell of Dell computers decided to custom prepare computers according to the specific needs and wants of every single customer, he was building his business on micro-segmentation.

When one slices the market into larger groups that represent a substantial number of people, we refer to **normal segmentation**. When one divides the market into very large segments, or even considers the market as one single segment, we call this **macro-segmentation**.

Marketers segment markets based on a few assumptions, in particular that:

1. Not all customers are alike,
2. Some groups of customers share some similarities,
3. One can choose a portion of the entire market to focus on,
4. This allows for an efficient marketing strategy.

---

[5] There are various models of segmentation, such as the VALS (values, attitudes, and lifestyles) system, which focuses on motivation and resources.

Segmentation rests on four pillars, which any marketing expert knows by heart. We divide the four pillars into two blocks: The first block, which includes the socio-demographic and the geographic pillars, treats hard data. (Both pillars, by the way, rest on data that are easily available and accessible in government databases.) These data are objective, fact based, and verifiable. Some marketing books narrow the first pillar to demographics, but, in fact, it is better to talk of socio-demographic information. (We consider this an "arguable error.") In all marketing studies, this includes sex (gender), age, and revenue. This is because purchasing behaviors vary widely depending on sex, age, and revenue, even though some steadfast marketers pretend men and women behave the same way, or that older people need the same products as younger ones. We can add more information (the "socio" part of socio-demographic) in some marketing research, where justified to do so and when the country where the research takes place accepts it. Such elements relate to culture at large or elements thereof, such as language, religion, ethnicity, family structure (marital status, number of children), as well as life stages, generations (e.g., Generation X), and so forth. Some marketing books mix socio-demographic and geographic criteria but, in fact, these are vastly different. (We consider this an "arguable error.") You can have the same socio-demographics in two completely different geographic areas; conversely, you can have different socio-demographics within the same, well-defined geographic area. Some books also posit the benefits sought by customers fall under segmentation criteria, but this contradicts the claim (made in the same books) that customers seek value; if they do, they consider both benefits and sacrifices (often reduced to the notion of costs). Following that logic, we should really say that the segmentation criterion is the value sought; in fact, value sought depends on the four traditional segmentation criteria: socio-demographic, geographic, lifestyle, and behavior.

The geographic pillar refers to where the customers live. Large retail businesses, as well as large infrastructure projects, invest heavily in geographic marketing (or, put differently, in geo-marketing). One wants to make sure customers can get to the point of sale (or point of use), so companies have a keen interest to avoid natural and human-based obstacles (e.g., a river or a train overpass). They want to identify where competitors are, and be as near to the customers as possible. Some astute companies will limit the presence of competitors in the neighborhood of their stores by buying nearby retail spaces even if they do not use it. The goal is to drive potential customers to one, and only one, destination: their stores.

The second block of pillars, which includes psychographic and behavioral measures, uses subjective data that require research by marketing experts, and hence some collection of **primary data** (data that does not already exists) on top of **secondary data** (data that already exists). **Psychographics**, also called lifestyle, refers to a cluster of standard behaviors that consumers repeat over time. Generally, there are often correlations between these behaviors. For example, the Hells Angels are generally easily recognizable because of their lifestyle. We think of them as driving Harley Davidson motorbikes, wearing black leather jackets and pants, being active in certain areas of the economy, not being necessarily befriend the police force, and so on.

A storeowner who sells biking equipment would be wise to sell black leather gloves, for example, if the local Hells Angels chapter has set foot nearby. Occasionally, marketing experts buy data for this pillar of segmentation from specialized research firms (often at high costs) or else seek them through field investigations.

Behavioral segmentation deals with isolated behaviors. These occur occasionally, often randomly, and display no correlation with specific purchasing patterns. A customer who is upset with waiting too long at a bank counter and who throws a temper tantrum provides an example. Obviously, with respect to these two pillars of segmentation—psychographics and behavioral—marketing experts must gather the information themselves. They often focus on

1. The needs the customers have and their readiness to buy,
2. Their decision process (alone, in pairs/couples, or in groups),
3. The benefits they seek,
4. The usage they make of the products (e.g., regular or inventive ways), and
5. Their levels of loyalty (e.g., hardcore believers, split-prone individuals, shift-prone individuals, or switchers).

Companies invest vast amounts of money to detect behaviors (and attitudes, as we have defined them before) because they may reveal new trends, expose hidden needs, and outline opportunities.

The **four pillars of segmentation** define a segment. We achieve good and efficient segmentation by respecting the six following criteria (some books refer to four, but this is short-changing the effort). The segment must

1. Express a demand for the product the company wishes to market. This is usually assessed through a needs analysis.
2. Be accessible: There is no point in spending money on a marketing plan if the segment is not reachable, as if the distribution system is underdeveloped.
3. Be sizable: The market segment should be significant in terms of volume. Per capita profits must justify the investment to develop the business aimed at a particular segment.
4. Be profitable: Except for not-for-profit organizations, profit is of the essence. Hence, a good market is one the marketing expert estimates capable of generating significant profits.
5. Be heterogeneous *versus* other segments: Each segment must be as different as possible from the other segments the expert has identified in the market. This way, the segment will receive marketing messages with better understanding and acceptance.
6. Finally, be homogeneous within the segment: Within each particular segment, the people forming it must be as identical to one another as possible. This way, the expert can create a simple, effective message without having to conceive of various transpositions.

Table 2.1 Segmentation Tools

Type	Pillars	Criteria
Micro	Hard data	Existence of needs
Normal	Socio-demographic	Accessible
Macro	Geographic	Sizable
—	Soft data	Profitable
—	Psychographic	Intersegment heterogeneity
—	Behavioral	Intra-segment homogeneity

We summarize segmentation in Table 2.1.

Obviously, project promoters will go at lengths to convince potential investors that their projects meet the six above-mentioned criteria, which apply to all segments and regardless of the kind of pillar under consideration. From a marketing feasibility point of view, being short on any one of these conditions is an indication that the project is precarious.

CHAPTER 2, CLASS EXERCISE #2:

Have fun working with segments; for example, find a wide variety of products and brands the same segment uses.

PROPOSED QUESTIONS FOR DEBATE:

3. Is segmentation a form of discrimination?
4. Is segmentation culture sensitive, or are market segments identical all over the world?

A typical way of illustrating segments in a simple format is with a pie chart, for example.

2.2.3 Positioning (and Competition)

Positioning is the art of understanding how a product (project, brand image, etc.) fares compared to its competitors. One does not segment products; marketers segregate products according, in particular, to positional maps. In the car industry, categories of products (and not segments) are typically luxury, sport, economy, family, minivan, utility, trucks, and increasingly, alternative-energy vehicles.[6]

---

[6] Segments would be, for example, value seekers, families, singles, high-income earners, and business owners.

Note that we can position people if, and only if, we treat them as a product. For example, marketing experts will develop a political campaign by positioning a political figure along, say, two axes: young and social. It is not the person the expert positions but the image of the person. The politician will have to dress somewhat casually, be seen or photographed running a marathon or meeting a group of seniors, and so forth. Again, talking about positioning people as people or segmenting products is an "arguable error," which (alas!) is often made by marketing people themselves.

Simple definition: **Positioning** is to products what segmentation is to people.

There are three kinds of positioning tools. In a **Cartesian map** with two orthogonal (perpendicular) axes, each axis stretches between two extremes: few ↔ many, weak ↔ strong, affordable ↔ expensive, cheap ↔ high quality, stinky ↔ pleasantly aromatic, etc. Each axis expresses a chosen attribute pertaining to the product under investigation, and the X- and Y-axis attributes are, preferably, as contrasting as possible.[7] How to determine which attributes to choose follows the process we describe below.

First, researchers draw a random list of all attributes that come to mind regarding the product. Brainstorming is of the essence; there can be 20 or 100 attributes, and they may be as weird as possible; it does not matter. Next, the researchers hierarchically sort the attributes the best they can, often intuitively, and often with the help of focus groups or market data analyses. Let us take an example: the car Joe keeps in his garage at all times (either because he can't drive without exceeding speed limits or because of snow), his flashy and expensive Lamborghini Urus.

A list of random attributes could be as follows:

Red	Sport	Performance	Roads	Fun
Expensive	Elegant	V12	Classy	Innovation
Fast	Sexy	Design	Macho	Luxury
Italian	Race	Atypical	Rich	Etc.

---

[7] We deal here with concrete attributes, sometimes called search attributes as opposed to experience attributes (which are attributes the consumers discover after or during the consumption of the good; e.g., how truly formidable the vacation was at a resort) or credence attributes (which rest on trust as they can hardly be measured; e.g., a doctor's assessment of your health condition).

Say we then create a hierarchy of ten attributes, as follows[8]:

1. Red	2. Sport	3. Performance	4. Roads	5. Fun
6. Expensive	7. Elegant	8. V12	9. Classy	10. Innovation

We ignore the remaining attributes for purposes of simplification. The next step consists in putting them on a map and linking together the various attributes, and putting the number of links besides the attributes, as follows (Figure 2.2).

According to this example, the attributes that come up most often (and hence assumed to occupy the customers' minds and hearts) are classy, performance, and sport.

These attributes will be useful later on, when we determine how to create a promise. Here, let us place them on a positioning map, with one attribute on the X-axis and another on the Y-axis (using expensive as an example here). We have the following positioning maps (Table 2.2).

By convention, we put the higher value of the X-axis to the right and the higher value of the Y-axis at the top. Products in the same quadrants as ours share **points-of-parity** (of resemblance), while products in other quadrants have

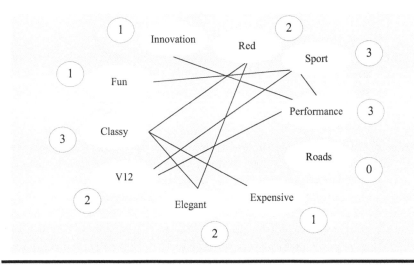

Figure 2.2   A perceptual map with a product's attributes.

Note: Perceptual maps are useful tools that allow for a rapid understanding of all elements at play in a given marketing situation.

---

[8] In reality, for cars, attributes that customers regularly consider are price, size, performance, interior, styling, service quality (including regular maintenance), safety quality, dealership proximity, and residual value.

Table 2.2   Examples of Positioning

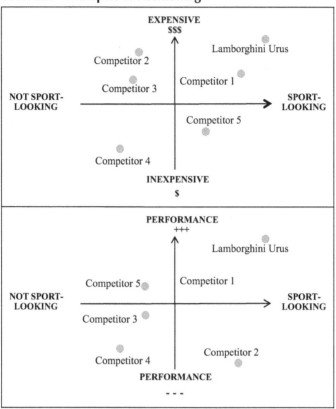

more **points-of-difference**. The more our product is positioned away from the competitors', the better it fares on key attributes that have been chosen, and the stronger its **competitive advantage** (the more value it offers). Sources of competitive advantage include attractiveness of the innovation, price leadership, band recognition, and of course, functionality and design. This easily applies to a project's deliverable as well.

Note we can create all kinds of positioning maps by using different combinations of attributes; it is up to marketing experts to determine what pairs of attributes are most useful given the coveted market. Note, also, that we could complete each graph by inserting names of the closest competitive products. This is what positioning is all about: finding out how a product fares against competition according to sets of attributes, whether they relate to, say, design, functionally, or price (or overall, to value). Doing so usually permits us to see where fewer competitors exist and, hence, where opportunities for growth lie.

Positioning maps can go into detail, and experts may even use them when comparing products that belong to the same company or the same family of products.

Rule of Thumb: A great adequacy occurs when the attributes of a product match the attributes of the designated market segment (the target market).

It is possible to create spreadsheets dividing the same attribute at various levels. Let us take a faucet as an example. In the rows, you would have faucets that are very expensive, affordable, and low cost. In the columns, you would have faucets used in the kitchen, the guest bathroom, bedroom ensuites, the laundry room, the main bathroom, etc. In doing so, you would also put the various kinds of faucets you offer as well as the faucets competitors offer. You would then be able to see where there is a hole, an opportunity. For example, you may realize that no one currently offers an inexpensive faucet for a master bedroom ensuite. Maybe there is a good reason for this; maybe there would not be enough demand. But, at least, such a spreadsheet would reveal all potential opportunities that exist given the two chosen attributes: price and room type.

We can make positioning more complex by using more than two attributes at once. In that case, we resort to **multi-criteria analysis**. It is customary to assign a weight to each attribute in order to determine which, at last, are primordial in the attempt to gain the customers' acceptance and loyalty. In fact, this is exactly what we did with the proposed Equation 1.1 on the decision to buy: *DC (decision to buy and consume) = f (β_1 Substitutes, β_2·Type, β_3 Utility, β_4 Budget, β_5 Urgency, β_6 Response, β_7·Opportunity)*.

Table 2.3 provides an example of this technique.

The number in the second row expresses the weight marketers assign to the attributes, which they can harmonize with the perceptual map done earlier. Here, the scale is from zero to ten, with ten being the most desirable attribute. As one can observe, after due consideration, the analysts deemed that "roads" (first row) did not deserve attention and they quantified this attribute by valuing it at zero (second row). The marketers can formulate a regression function that takes into account the weight against various criteria. There is more to multi-criteria than simply this chart. In particular, analysts usually add an additional line according to cost of manufacturing the product (third row). Performance cost, linked to the V12 attribute (valued here at seven), requires expansive engineering studies, as does innovation. Elegant and classy belong to design work,

Table 2.3 Simplified Multi-Criteria Analysis

Red	Sport	Performance	Roads	Fun	Expensive	Elegant	V12	Classy	Innovation
7	10	10	0	3	3	7	8	10	3
1	0	7	0	0	0	3	7	3	7

and the project managers estimate here that this aspect of the product development is less demanding financially (valued at three each). As one can see, the manager must take many criteria into account, hence the name multi-criteria analysis. Performance is certainly a key attribute (valued at 10), but it comes at a high cost (7). In the present example, innovation is not an important criterion for the consumers (3), but the cost of innovation is equally high (7). What would be best? A project to build a product using existing technology but offering a better design, or investing heavily in innovation to boost performance and break through the price ceiling. Consumers (mostly wealthy) will pay regardless and happily so. Innovation and performance are certainly correlated, so playing with one will affect the other. Multi-criteria analyses can become quite complex. In this particular case, marketing and project managers want to determine the best combination of attributes given the costs.

Suppose a project has a budget of USD 100,000. The project managers must complete it within one year, and they must respect certain norms of quality. Managers must often decide on some trade-offs: They may decide to postpone the delivery date by one month, or replace one material planned for the product being developed with a cheaper one because the original material is no longer available (but the replacement cannot be delivered until two weeks later, and this will be done at a premium cost). In the end, project managers will surely complete projects, but perhaps not as originally planned. Ideally, the three attributes—timeline, budget, and norms of quality—are at the top of project managers' list of priorities. Ideally, they should be on equal footing, but circumstances sometimes force marketing and project managers to sacrifice one for the benefit of the other, hoping clients will receive the same level of utility (and experience the same level of satisfaction) in the end.

What the managers have done is list the criteria, thought of two scenarios, and assigned a weight to each. (In the example, all three criteria are equally important, as is the prevailing case for the vast majority of projects.) (Table 2.4)

If we transpose these three conditions into the evaluation of marketing feasibility for projects, the question becomes: Can we deliver and implement a marketing

Table 2.4 Multi-Scenario Decision-Making

	Weight	Scenario 1	Scenario 2
Deadline	10	Keep current plan	Extend the delivery date
Budget	10	Boosted to order the needed part from another manufacturer, at a premium cost	Slightly changed
Quality	10	Higher	Lower
Final choice		This scenario?	Or this scenario?

plan for the project that we can deliver on time to promote the project adequately, that meets a preset budget, and that will be of a high-enough quality to convince prospect customers? When project promoters present their projects to investors, this is usually what they do—ask these kinds of multi-criteria questions, sometimes without even realizing it. They sell their projects, and they try to show they can market them in a way worthwhile bringing them to completion.

These examples demonstrate how important it is to understand product attributes and to position products accordingly.

Another key concept with respect to attributes is that of primacy. **Primary attributes** are things customers absolutely want, and not having them will discourage these would-be clients from buying the product. Primary attributes are *sine qua non* conditions for purchasing the products. In Table 2.3, the marketer quantified them at ten (e.g., sport, performance, and class). In projects, *sine qua non* conditions are meeting deadlines, respecting the budget, and obliging preset quality standards. **Secondary attributes** are nice to have, but are not essential to the decision to buy the products. In the example above, red is a nice color to have, very nice indeed, but customers may be just as equally happy buying a yellow Lamborghini, even though their mind was originally set on a red one.

Customers resort to positioning when deciding on which products to buy. They position competing products against the coveted products and debate as to which attribute is most important or is a must-have.

So far, we have positioned products with respect to competitors and attributes. We can also position products according to where they are on their **lifecycle**: introduction, growth, maturity, or decline (see Figure 2.3).[9]

Products at the end of the cycle will likely face an exit strategy. Projects, too, have their lifecycles, but managers treat these differently than product lifecycles. (We will review this in more detail in Chapter 3 on projects.) There are a number of entry and exit strategies that marketing books discuss at length; many use military terms such as "encirclement."

Entry strategy are plans designed to enter a market that is new (or not) and with a new product (or not), or a combination thereof.[10] The standard framework for such strategy is given in Table 2.5.

By definition, with projects (being the process of materializing a product), only two situations are possible, which we put on the left side of Table 2.5. An engineering company may develop a new product in a new market (e.g., the hospital in the Gonaives after Haiti's devastating earthquake; see Appendix 1), or else simply develop a new product for the existing market, such as in the example of the multifunctional center in Québec City. (See Appendix 1.)

All projects, being new, necessarily take into consideration entry strategies. Marketing science has it that different such strategies involve the 4Ps of

---

[9] PMBOK 6, 2017.
[10] For more on this, see the Ansoff matrix.

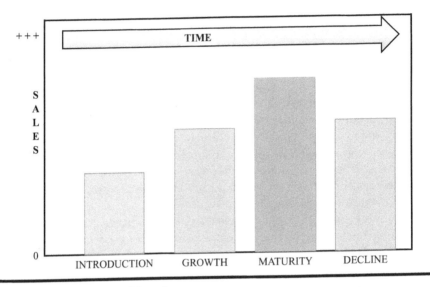

Figure 2.3 A product's lifecycle.

Note: The concept of product lifecycle is instrumental in marketing. Each phase tells the story of competitors, consumers' reactions, and profit potential. Obviously, if sales reach a peak and then start declining, consumers have lost interest and competitors are eating away at the acquired market share. A project for a new product may be advisable.

Table 2.5 Marketing Entry Strategies

	New Product	*Existing Product*
New market	Project	
Existing market	Project	

marketing: product, price, promotion, and place. As an example using price, marketers may introduce a new product into the market at a high price in order to maximize margins to the detriment of volume, benefiting from opinion leaders' recommendations; this was the case for the iPad in the early 2000s. This theoretically allows companies to recoup R&D expenses faster. Experts call this **skimming**.

Another entry strategy is to sell the product at a lower price and hope for a large volume of sales, a strategy called **market penetration**. Either way, marketing experts work in tandem with project promoters to determine which is the best pricing strategy in order to make the project not only feasible, but also sustainable. (The two terms are different, but people tend to confuse them, an "arguable error.") Feasible only means project managers have the forces and means of actually materializing the envisioned product (deliverable); sustainable (or in a large sense, viable or profitable) refers to what happens after the managers have completed the project. As discussed in the general introduction, feasibility is about upstream thinking.

Strategies for international market entry include direct and indirect investments, joint ventures or other types of partnerships (often the case in large-scale projects), licensing, and franchising. International marketers favor one of two approaches. They either present a uniform promotional campaign to world markets (e.g., to advertise the next Olympic Games) or tailor the message to each market by adapting products to particularities of each country (e.g., as is the case for many food outlets, like McDonald's offering the McSpicy Paneer in India, the Le Croque McDo in France, and the Mega Teriyaki Burger in Japan).

Once the project has actually come to term and has become a product, the marketing of it belongs solely to the marketer. It happens, however, that project managers cannot bring their projects to completion, the Olympic Stadium in Montréal or the Sagrada Familia in Barcelona being prime examples. In such cases, the project is no longer a project *per se*, because all projects have a deadline; here, there are no fixed deadlines, or they have deadlines that are completely past what the promoters originally planned. In such cases, marketing experts will consider growth, or survival or exit strategies. In short, marketing and project managers are jointly concerned with a product that is at the beginning of its lifecycle, at the limit where the project ends and turns into an operation. Initially, marketing feasibility analysis serves to determine not what the marketing plan for the project should be, but rather if that plan is feasible. At first, it is not the marketing plan for the product excavated out of the project that marketing and project managers consider, but rather the marketing plan of the project itself. This distinction is important, although people sometimes confuse the two challenges. Of course, there has to be a marketing plan for the deliverable. As discussed in the introduction, the plan is twofold: marketing the project to entice funders to finance it and marketing the deliverable that comes out of the project. The promoters must prepare for both marketing efforts.

CHAPTER 2, CLASS EXERCISE #3:

Choose a product you dream of buying, but cannot afford right now. Draw a Cartesian map showing the various options you have to approximate this product, using attributes dearest to your heart.

PROPOSED QUESTIONS FOR DEBATE:

1. Can market opportunities be identified using positioning?
2. Should innovative products fill all "voids" in the market?

2.2.4 Targeting and the 4Ps

Targeting is the next step after segmentation and positioning. Remember, however, that there are many iterations between the six components of marketing strategy. The tools used for segmentation, as we have seen them, are listed in Table 2.1. Marketing experts use these segmentation tools, such as the four pillars, and then the positioning tools, such as perceptual maps, Cartesian positioning maps, and multi-criteria analyses. In the realm of targeting, the number of arrows used by a marketer-archer who targets one, or many, market segments is four. More to the point, a key concept in marketing is, of course, the **4Ps:** product, price, promotion, and place or points of distribution (e.g., website, a store, a fair). We could say a lot about each of the 4Ps, but this book only addresses them briefly.

Product is, evidently, what we have already discussed: products, services, ideas (messages/brands), experiences, and projects. Prices relate to the actual cost of manufacturing the product and its sales price to customers. Promotion encompasses activities like trade fairs, as well as advertising (usually paid for by the suppliers) and publicity (not necessarily paid for by the suppliers; think of bad publicity, for example). As mentioned, place refers to points of distribution, which marketers segregate in a wide variety of ways. Evidently, agents along the distribution chain store and move products, but they also often facilitate transactions and logistics. Not to mention they can be a great source of market intelligence. These 4Ps are an integral part of all marketing plans, as we shall soon discover.

> Simple definition: **Targeting** is about choosing a mixed marketing strategy to best reach the intended clientele.

For the sake of linking marketing and project management, however, it is useful to add right away to these 4Ps of marketing the 4Ps of project management, which we will discuss in detail in the upcoming chapter. Figure 2.4 exemplifies this link.

To put it simply, marketing efforts, in the framework of projects, must consider needs like opportunity (expressed as a promise), as well as people. Marketers work not only with the traditional 4Ps but also with the 4Ps of project management. Project managers would be well advised to define their projects methodically (product), to understand their value in the eyes of the customers (selling price in particular), to foresee how they will promote them (if only because media often follows up on major project developments), and to determine what place they will occupy in the community where these products will be operationalized. Project managers make use of the 4Ps of marketing.

Targeting is identifying the core market segment where marketing activity will take place by way of establishing product, price, promotion, and distribution (place) strategies—the four arrows of the marketer-archer. The goal is to capture the mind, heart, and wallets of the prospect customers (see Figure 2.5).

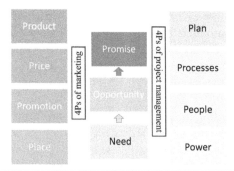

Figure 2.4   The 4Ps of marketing and project management, and their links.

Note: People are all the stakeholders whom any marketing plan should consider, such as customers and salespeople, for example. Promise is the essence of projects, and of any offering: A project is a promise (see Chapter 3)[11] and a product is a promise, that of fulfilling the needs customers experience as a void and that they must respond to in order to re-establish homeostasis (stability).

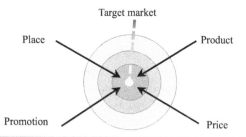

Figure 2.5   Targeting with the 4Ps as arrows of sorts.

Note: Few marketing books relate the 4Ps of marketing to targeting, but this is where they belong. There is no point in offering a product at the wrong price to a set of uninterested customers. The best example is that of women's hygiene products: Clearly, they target a specific market.

Marketing experts have written countless books on each one of these 4Ps (together called the **marketing mix**). It is important to note that they do not exist in a vacuum, and marketing managers use them as targeting tools; strategically, targeting follows segmentation and positioning. Let us discuss briefly each of the 4Ps of marketing so we can have an overview in the context of project management.

We mentioned products previously; to recap, a project is generally a product coupled with some form of service; this is especially true of large projects such as infrastructures, like roads, that will service a large portion of the population.

---

[11] A warranty is also a promise.

Governments spend money to build bridges and roads, and people use them. Bridges are concrete products turned into operations once the projects are completed. They provide the electors with a service: moving from Point A (home) to Point B (work), for example. Promoters design many projects to offer the end users a unique experience. (See the case of the Montréal International District [MID], Appendix 1.)

Prices are set according to various modalities. Some forms of pricing are illegal, such as **predatory pricing** (when sellers set their prices solely for eliminating competition). This is illegal because we base our economy on the ideal of perfect competition, by which consumers get what they want, when they want, at a reasonable price. A seller (company) who has managed to run its competitors out of business ends up in a monopoly. This allows the seller to set prices the way it sees fit, and not in the best interest of the consumers, as monopolies make inefficient use of resources, as mentioned before. Collusion is also illegal, although regulators regularly catch, and fine, companies red-handed. Often, the fines are ridiculously low compared to the profits the collusive sellers pocket.

Legal pricing includes, among other options:

1. Adding total production costs and profit margins, where margins can be calculated in percentages (the sellers want 10% above cost) or fixed amounts (e.g., they want one euro above the cost of every unit),
2. Using psychological price-set points (£4.99 instead of £5.00),[12]
3. Adjusting prices against that of competitors without starting a price war (price wars are generally detrimental to all sellers),
4. Implementing promotional prices (e.g., two for the price of one),
5. Determining prices based on **elasticity**. (A very elastic price would be one where a small change—increase or decrease—in price would lead to a very large change in the quantity demanded—negative or positive.)
6. Setting a price floor (the lowest sellers will accept, such as when negotiating the sale of a vehicle at a dealership) or a **price ceiling** above which the product becomes irrationally expensive in the eyes of the consumers.

As for customers, they can use a variety of references to determine whether a price is fair or not: last price paid, comparisons with competitors, historical pricing (e.g., in the real estate business), lowest expected price, highest expected price, etc.

Making a profit is not necessarily the only objective in fixing a price. Marketers may have other reasons, such as increasing sales (often by actually reducing prices);

---

[12] Some marketing books argue that psychological pricing makes it easier for customers to buy products, but this statement is an "arguable error." In fact, when customers have a basket of products, each priced at C$ X.99, for example, it is much harder to calculate the total than if the products are priced at C$ X.00. Hence, the policy of C$ X.99 confuses customers, who then buys without really calculating their benefits and sacrifices—a dream for sellers.

increasing market share (to the detriment of competitors); meeting customers' expectations and their ability to pay, fulfilling a social mission; and finally, improving the brands' image (some customers often interpret a higher price as a guarantee of quality).

In the case of projects, promoters have set the budget in advance and have generally and carefully determined the end users' fees; otherwise, they would have no convincing arguments to present to potential investors. The mere fact of deciding to embark on a project means there is a demand and that consumers will use it and generate profits for the operation. At least, this is the goal, although it is not always achievable.

As for promotion, we should mention three important concepts: coder-decoder, culture, and placement. **Promotion** (and/or publicity) is about highlighting the primary attributes identified during the positioning process in order to convince would-be consumers that our products will bring them **satisfaction** (more benefits than sacrifices in fulfilling their needs); that is, our product will offer at least enough utility to meet their needs or wants, and, if lucky, even more. Promotion is ensuring that the minds, hearts, and wallets of the consumers remember at all times that our products exist (not that our products are the ones which stand atop others, which is the job of differentiation) and is waiting for them, so to speak. Effective communication invites consumers to action; most live in a state of relative inertia unless the need becomes pressing. To paraphrase Henri Troyat, they are often buried in the amorphous paste of daily life![13]

The first key concept associated with such communication effort is that of the **coder-decoder**. In essence, marketers code the attributes by using terms to which the consumers will easily relate. Instead of saying "the pill will stop you from farting in public," it may be better to say "The pill will protect you from a possible source of embarrassment at the next beauty contest." The message is sent and the receivers—the prospect clients—receive, perceive, and decode the message after it has gone through various filters, such as cultural beliefs and cognitive biases. We discuss perception in the section on differentiation.

If you do like Ford did in the 1970s, when it launched a marketing campaign in Latin America to sell its Nova car, arguing "nova" meant "new" in Italian, you might be in for a surprise. Potential buyers decoded the term in their native language, Spanish: As we discussed, "no va" means that the car does not move.

During the communication process, noise from various sources of interference will blur the message. Some experts estimate that, in certain countries, consumers receive no less than 5,000 messages a day, and some are bound to interfere with the one message you are trying to send. The illustration of this well-known communication model is as follows (Figure 2.6).

This model supposes that market agents have an interest to communicate, a motivation. In other words, customers have realized they have a need they rightfully

---

[13] Troyat, H. (1951), *La tête sur les épaules*. Paris, Librairie Plon, p. 131. Our translation.

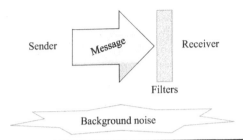

Figure 2.6 A communication model.

Note: This communication model is a standard in marketing theory, although it takes various forms and may adopt various levels of complexity.

wish to fulfill; they thus become receptive to messages that invite them to find a solution to their condition. Their receptivity increases, and they pay full attention to the message and its inherent promise. They dig into their memories to better assess the information they receive and may seek other information from their environment, including from the sellers they face. They interpret the overall message (the content, but also the way marketers present them). Through this entire process, they learn about themselves, the product, and their environment.[14]

In our contemporary world, a number of problems plague communication networks that are worth mentioning:

1. More and more receivers of messages feel stretched between various sources of information, of which the credibility is often dubious (e.g., TV, websites);
2. Communication is being amputated of some fundamental elements: e-messages often imply people do not pay attention to sentence structure, grammar, syntax, or verb tense, or do not keep a common or logical thread among their ideas;
3. Electronic communications also lack many components that give sense to a message, including emotions, and are thus subject to misunderstanding;
4. Ease is the rule of (up and down) thumb, so to speak. People make judgments about everything without realizing it, yet these judgments are biases that infest their thinking and may lead them to the wrong decision; and finally,[15]
5. Speed is of the essence. Various reports estimate that Internet users do not spend more than three seconds per page before zapping.

---

[14] For each of these steps, there are a number of theories. For example, there are theories on perception and learning (such as those promoted by the behaviorist school or by the cognitive theory proponents).

[15] The expert will easily detect the judgmental approaches and biases of people answering the exercises provided during our seminars, designed in particular to help these people to become cognizant of them in order to produce fair and objective marketing feasibility analyses.

Given these dreadful realities, consumers actually become less educated, less prone to making sound judgments, more irritable, and less willing to hear the message that, say, the project promoters want to convey. Feasibility analysts should be aware of this, because they themselves are part of these realities.

All **filters** affect the flow of communication and are, in essence, cultural. There are a number of definitions of culture. Roughly speaking, a **culture** is a set of sustained elements bounded within a territory. These elements include architecture and buildings, artwork, biases (e.g., anti-Semitism in certain countries and at certain epochs in history), clothing, exemplary humans (e.g., an athlete), food, language, religion, social habits, some infrastructures, symbols, legal systems (e.g., Common Law for Commonwealth countries, Napoleon Code for French countries), traditions, values, and so forth. Cultures change over time. For example, in the 1960s, experts knew Québec for harboring six fundamental groups of traits, each containing six sub-traits (including Catholicism). Nowadays, almost none of these traits prevail. The Québécois have moved, in some sixty years, from a French Catholic society (very much French in its way of thinking and behaving) to being truly North American, with the only difference between them and English-speaking Americans or Canadians being that they speak French. Culture is shared horizontally (among peers) and vertically (down the generations). Culture is enduring yet flexible, and it is vital to an individual's identity. As such, culture influences all filters in the communication path we drew in Figure 2.6. To take the Québec example again, publicists know well they cannot use the same ads in French and English Canada; the sense of humor is sparklingly different between the two groups and the Québécois very much dislike seeing a TV ad where the actor is trying to replace the original discourse from English to French. They long for a genuine sense of culture, including in their food—something they have kept from their French origins. As another example, let us take McDonalds. In the 1970s, the company tried to market its hamburgers mixing cheese and meat to the Hebraic society, whose religious beliefs forbid the mixing of milk products and meat. As for Disney Paris, a huge project undertaking, it had a real difficult time in launching until it adjusted its procedures—most notably its meals and eating areas—to match the expectations of its clientele (Europeans) who don't eat like Americans. We can say that culture, in terms of project management, is formed from the 4Ps: plan, processes, people, and power.

The third concept of importance with respect to promotion is that of **placement**. The messages sellers wish to transfer to their target audiences use a specific medium. It can be word-of-mouth or a media (TV, radio, journals, magazines, newspapers, billboards, pamphlets, street pavements, one's tattooed body or clothes, and so forth). Placing ads in the media is often expensive and requires sellers to decide when (e.g., day, time of day, time of year) the message should be conveyed and where (e.g., which medium or geographic location). Marketers must also decide how many times (five times a day, three times a week) the message will be conveyed, in what way (humorous, or appealing to intellect, emotions, or attitude), to who (the target market, of course), how (the actual media), and why (to encourage sales, trigger interest, incite behaviors, etc.). Expert marketers believe that a message must be heard at least five times from at least three different media over one week, and repeatedly so, in order for it to sink into the targeted consumers' minds, hearts, and wallets. It is a complex task and, again, academics have written many books on the subject.

Broadly speaking, marketers resort to three ways to communicate their messages:

1. Wide publicity, price promotions (price reductions, special events, coupons, draws, contests, sampling, etc.), and public relations (e.g., press release, sponsorship);
2. Personalized, targeted messages sent to a specific segment, relying on databases, direct marketing, and personal selling; and
3. Multifaceted approaches, such as is the case with social media or viral marketing, where many sources emit many similar messages to many different publics.

Publicists attract consumers' attention by resorting to their intellect, their emotions, or their urge to act (open their wallets), or a mix thereof. They tease, compare, and demonstrate; they use humor, romance, sex, and sometimes fear (although the use of fear usually leads to a reverse effect). They invite celebrities and reinforce their messages with jingles and slogans.

Projects do not accommodate well the usual marketing tricks that include weak and sneaky promotions. Project managers best promote their projects with accurate information, by highlighting their primary attributes and benefits, and by being clear on the cost the end user should expect. In the case of cities wanting to host the next Olympic Games, a promotional campaign targets first the promoters and decision-makers and, near completion, the end users. This is usually the case in all projects, as there are generally two promotional phases: one to convince promoters and the other for the end users once the project nears completion (see Figure 2.7).

Place is the last of the 4Ps, after product, price, and promotion. Two concepts are primordial, that of the **length** of the distribution chain[16] and that of the **depth** of the product line. The length of distribution refers to the number of market players that exist between the manufacturer and the final seller (e.g., the retailer). A long chain means there are many intermediaries, such as brokers, distributors, and warehouses, each taking their own commission and thus each increasing the cost of the product by the time it reaches the shelves (whether virtual or real). The depth of the product line is the number of products that pertain to a certain category. For example, a café may have five different sorts of coffees

---

[16] For more, read about the concept of supply chain management, on which experts have written many books.

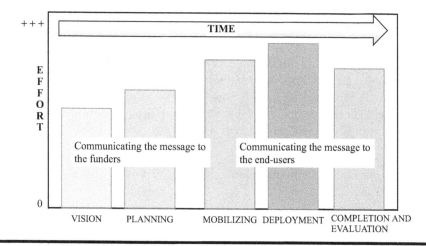

Figure 2.7   A typical communication pattern in the lifecycle of a project.

Note: Lifecycles dictate various marketing and managerial strategies, including with respect to investments and, of course, communication.

(cappuccino, espresso, etc.), each with a different country of origin (e.g., Brazil, Columbia, Nigeria). A point of sale (the café) may have, for example, a combination of products from various countries for a total of twelve SKUs (stock keeping units), whereas its competitors will welcome their customers with a choice of 50 different combinations; the latter has a deeper line.

One can see how the 4Ps of marketing cover quite a wide range of strategic issues, and how they will affect demand (quantity demanded, as previously seen). A product that answers a need, at a good or reasonable price, and of which consumers are aware through promotion, or is easily accessible through proper distribution, has an excellent chance at success. If the product targets the wrong market segment, or else occupies the wrong category of products or sets of attributes, it will likely fail.

CHAPTER 2, CLASS EXERCISE #4:

Choose a product you use regularly. Investigate where, and when, you are exposed to information about it, both formally (ads) and informally (friends, product placement in movies).

PROPOSED QUESTIONS FOR DEBATE:

1. Should movies promote brands?
2. Do you feel you are overexposed to advertising and promotion for certain products or brands?

2.2.5 Differentiation (and Branding)

Differentiation is about ensuring consumers recognize our products that sit on a shelf that harbors hundreds of products offering the same utility. Segmentation and positioning are forms of differentiation. Marketing managers use the products' appeal and the messages they carry (through the 4Ps, including, of course, promotion) to highlight their key points of differentiation. They distance themselves from the points of similarity (parity) that exist with other products.

> Simple definition: **Differentiation** is making sure we stand out against competitors.

In terms of projects, differentiation is a given. About all creators of a major project pride themselves on how their products, which are a deliverable of the project, will be memorable, if not life-changing altogether. The project acts as the brand name. Attachment develops as users get acquainted with it, feel safe with it, recognize its utility, benefit from it in as many aspects of their lives as possible (the more, the more memorable), share it with others (eventually building a community), and recognize its symbolic value within their culture. The Eiffel Tower is a perfect example. The engineer who conceived it, Mr. Eiffel himself, did so as a pure project—it was not even supposed to have any meaning past the Paris Exhibition of 1876. Yet, every year, France welcomes more tourists than it has citizens, most eager to marry under the tower, to climb it, or to do selfies with it in the background. It differentiates Paris from any other city in the world. (Well done, Mr. Eiffel!)

Let us set a product that generates little apprehension, one that has a great reputation and that consumers know for respecting its promise (providing the expected utility). What is going to happen then? Well, consumers will trust the product and make sure they can identify it easily when submerged in a sea of similar products. To recognize the product, consumers use clues such as colors, logos, slogans, and so forth. In summary, the shortest way to winning the consumers and enabling them to identify our product is through the creation of a brand. A **brand** is the identity of the product, its given name.

When a product becomes a brand, it transcends its mere utility-driven entity. Properly empowered, it can become part of the family. People have used some products for generations, and they indeed considered them part of the family; Nestlé Quick has delighted kids for decades. What we are saying here is that differentiation goes both ways, an important remark some marketing books tend to forget. Not only does the product need to be differentiated, but also consumers have to have their eyes opened to see the difference.

Hence, differentiation tactics appeal to many concepts used in marketing perception theories, including biases, heuristics, and graphic illusions. To make our point clear, let us consider an old movie in which a car is moving forward.

To the viewers, however, the wheels move backward because of a stroboscopic effect. The filmmaker communicates the message the car is moving forward, but the receivers, the onlookers, think it may actually be moving backward when they focus on the wheels' movement. It is hard to convince customers of the value of our products if we send conflicting messages. Yet, if we can use the stroboscopic effect to our advantage, consumers will remember us. **Perceptual and mnemonic tactics** are two other key tools marketers use to ensure differentiation, besides brand creation. Key perceptual effects managers can use to differentiate their products include the so-called recency effect, among others. (We do not cover them here, as they have no bearing on project management.) Examples of ads that differentiate a product and invite consumers to pause and seek to understand its content include those of United Colors of Benetton, which everybody knows in Canada, for example.

Many marketing books regretfully treat these elements without incorporating them as part of the overall marketing strategy or as part of the differentiation effort (this is an "arguable error"). Differentiation does not exist in a vacuum, and neither do its perceptual and mnemonic tools: Using them wisely ensures consumers segregate products among those of our competition. Indeed, perceptual and mnemonic tricks can serve to magnify the attributes that consumers would otherwise easily confound with other products. Consumers assess a project's products and deliverables before they buy, while they buy, once used, and even after (e.g., when disposing of packaging). They rely on such criteria as product quality, of course, but also on availability, technical assistance (e.g., in the computer industry), warranty, follow-up services (e.g., in the car industry), and the overall buying experience.

The best argument in favor of putting perception effects into the differentiation component of the marketing strategy comes from brand's slogans. **Slogans** are short lines aimed at making sure customers remember the brand in all circumstances. The brand name has to stick into the hearts, minds, and wallets of the customers. To achieve this, marketers and ad designers often build their slogans based on astute effective perception effects. Mantras are often short, punchy, clear, and express at least one key attribute of the product; they preferably promote action, may use onomatopoeia (because customers find that amusing and thus entertaining and memorable), and avoid ambiguity. Preferably, they are specific to the product, or category of products, they represent. Great slogans are inspiring, cross generations, and often speak to the three components of attitude: emotions (including the senses), cognitive (including symbolism), and conative (inviting action). Ideally, mantras that express an attitude are ones which their adopters feel proud to carry socially. The brand tells a story about how they feel, think, and act.

Here are some examples of slogans (Table 2.6).

Brands present a number of benefits, including

1. A warranty of products' quality;
2. A reduction of perceived risk favoring a positivity bias toward the product;
3. A simplified decision-making process;

Table 2.6 Examples of Slogans

Slogan	Company	Build-On or Problem
Notoriously Effective		
"Das Auto"	Volkswagen	Positive bias toward German car engineering. Differentiates from non-German cars.
"Just do it"	Nike	Promotes action
Dubious		
"Je le vaux"	L'Oréal	We suspect L'Oréal changed it to "parce que (vous le valez bien)" ("because you're worth it") because "vaux" sounds like "veau" (meaning veal).

    4. A reinforcement of the meaning of the product, making it more personal;

    5. An opportunity to become part of the customers' identity; and finally,

    6. A worldwide uniformity and hence reduction of marketing costs.

Other core concepts of interest include **brand management**, brand attachment, brand equity, and brand community.[17] Brand management is simply the action of managing the brand and with it, brand attachment, equity, and community.

    With respect to brand attachment, experts have long noticed consumers develop an emotional bond toward some specific brands. The brands mean something to them, something that surpasses a mere name or logo. Coke, which we discussed earlier, is an excellent example. There can be no voluntary attachment with the name in the presence of apprehension, neither can there be attachment if perceptual fault lines distort the message the product carries, or else when the product does not surface spontaneously in the minds of the consumers when a need emerges. In the exercise we proposed in the section "Positioning," we asked participants to identify spontaneously brands they use every day. The fact they were able to list these brands means that, somehow, the chosen products detached themselves from competitive products.

    For brands to have strong equity, they must be

    1. Culturally adapted,

    2. Likable,

    3. Meaningful,

    4. Memorable,

    5. Noticeable,

    6. Safe, and

    7. Trustworthy.

---

[17] Various brand management models exist such as the Brand Resonance Model.

They must also offer a superior value (functionality/utility and design, given costs) for a reasonable share of benefits and sacrifices, and finally, they must carry a **symbolic value**. Marketers achieve the latter, which is no small task, by associating brands with the customers' life components such as

1. Generations (e.g., Nike),
2. Historic times and places (e.g., New York City, the Big Apple),
3. Life stages (e.g., Jell-O),
4. Reference groups (e.g., Harley Davidson),
5. Social classes (e.g., Louis Vuitton),
6. Subcultures (e.g., Red Bull),
7. Tendencies (e.g., Christian Dior), and
8. Traditions and rituals (e.g., the old Sears home-delivered, much-anticipated catalog of products).

Brand equity is built by enforcing these qualities when selling the products (or projects), and by extending the product lines the brand contains (so-called brand extension) and/or its distribution points. Brands are the expression of the product's value, as we've defined it; technically, brand equity is the perception of a brand (intangible) which is operationalized by financial measurements of value. As discussed, they are not only the promise but also the vivid proof. Strong brands, including those associated with projects or projects themselves which are treated like a brand, create a positive bias in onlookers, promote loyalty, limit vulnerability to external forces (including market hazards and competition), justify higher margins, and reduce consumers' sensitivity to price changes (**elasticity**) and perceived risk.

There exist a number of international measures to assess the value of brands there are numerous listings available on the Internet, such as Forbes', that rank the top brands according to various criteria and over the years. They consider brands such as Apple, Microsoft, BMW, Toyota, and L'Oréal. These include sales and how easily customers associate with the product by spontaneously giving the name of the brand when asked for one randomly.

With respect to brands building community, marketers like to resort to **opinion leaders**, speakers, and advocates.

Opinion leaders are personalities who have a strong influence on the targeted audience because they are well known and respected for their competence, expertise, or talent. Marketers do not necessarily have to pay for their services; the mere fact they recommend, say, a book (like this one!), boosts its sales without any investment in promotion. For example, when Oprah Winfrey recommends a book, her followers flock the bookstores by the hundreds, sometimes by the thousands.

Community building through brands is a dream come true for marketers. Any customer belongs to groups of references, such as family, friends, colleagues, hobby-related groups, sports teams, or professional associations. Within these groups, people talk and voluntarily (or not) promote the brands they use and thus provide promotion free of charge to the brand owner. Starbucks uses this strategy in its own way. It does not publicize in the media because it does not need to; customers do it for them. They leave the store with a cup in their hand, a cup that displays the company's logo, and happily and unconsciously share the information with their immediate neighbors and, once at work, with their colleagues.

People talk about projects, especially big ones, like brands, and hence, marketers and project managers can treat them as such.

Overall, differentiation is about using the psychological map of customers, their perception mechanisms, and brand management (which includes slogan creation and the establishment of brand attachment, equity, and communities). By **perception mechanism**, we refer to the standard model of exposure to stimuli, attention (interest), interpretation (coding and decoding, with filters and biases), and retention.

Figure 2.8 illustrates how marketers use differentiation through the establishment of a strong brand (or projects, as end users often see large projects as brands in their own right). As can be seen, brand building is a long-term effort that heads toward customer loyalty (the last of the six components of marketing strategy).

CHAPTER 2, CLASS EXERCISE #5:

Identify a strong brand and discuss every one of its aspects based on Figure 2.8.

PROPOSED QUESTIONS FOR DEBATE:

1. Do brands occupy too much of consumers' time?
2. Would money be better spent doing less brand promotion and putting it in community development?

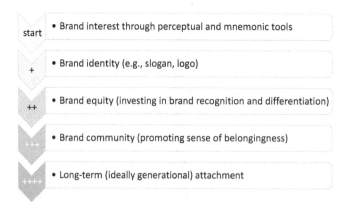

Figure 2.8 The building flow of a strong brand.

Note: Marketers build brands over time. In our digital age, brands can reach international fame within a few years, whereas this took decades a century ago.

2.2.6 Loyalty Building

Loyalty programs are a favorite tool to promote loyalty, just as special events carry a strong sense of belonging to the community, as proven, for example, by Bombardier Recreational Products' Can-Am Spyder gathering that takes place every year.[18] The goal is to make consumers proud of the products they consume, and to use these satisfied customers to bring more selected consumers onboard.[19]

> Simple definition: **Loyalty** is secured when clients are satisfied over the long term.

Loyalty programs constitute a reward for the efforts consumers have made in buying and using the product. Indeed, as previously discussed and contrary to common belief, consumers must spend energy and effort, on top of money, to acquire goods. The simple fact of going to a grocery store implies preparation, at least mentally: making a shopping list, deciding on a means of transportation to reach our favorite point of sale, knowing we will be waiting in line when we could be home watching our favorite sitcom, and then unpacking the goods when we get home or when they are delivered. (Grocery stores, and IKEA, should pay the consumers for their work!) To acknowledge this fact, marketers develop loyalty programs that come in various forms, such as memberships, exclusive client lists, travel points, and so forth. The success of these programs rests in the psychological framework we have seen earlier. Consumers must feel safe and secure (little or no perceived threat), should feel they can trust the product because it delivers what it says it will deliver (the expected utility), and because they feel they are treated fairly and the seller cooperates with them (e.g., by listening to their grievances and complaints).

The second tool of the loyalty component is therefore the **promise**. A promise is what the sellers would say to the buyers to comfort them in thinking they are making the right buying decision and should hence become or remain loyal.

A promise is what the project promoters, for example, tell end users about their project or deliverable (product). Marketers develop promises based on the primary attributes. In the case of the Lamborghini, seen earlier, we posited that the car had three attributes that stood out from the others: sport, performance, and class. The promise for the car, as customers perceive it, would therefore be: We promise you a classy sports car that brings you high performance. Transforming the promise into a slogan and you get something like "Top speed for top players."

---

[18] https://lam.can-am.brp.com/on-road/owners/spyder-blog/the-inside-scoop-on-the-7th-annual-owners-event.html. Accessed March 3, 2019.

[19] Loyalty ranks high among managers as a key indicator of their businesses, besides profitability, and competition.

Another terminology used for the promise is the **value proposition** (or initial value proposition as seen earlier). The question then becomes: What is the value? As discussed, some marketing books pretend marketing consists in bringing value to customers. That is not true and is an "arguable error"; one can well put money into an advertising campaign for a product that brings very little value to customers. The three core groups of attributes customers stick to, just about anywhere in the world, are design, functionality, and cost. Originally, Apple built its business on design (the all-in-one computer format) and functionality (the mouse and the icons). Design is the ability to please the customers' senses. Functionality is, in economic terms, utility, that is, the capacity to meet one's needs. Cost is a no-brainer, and we saw how it is the trade-off between benefits and sacrifices. Therefore, really, when we plotted quantity demanded against price in a previous figure, we actually only focused on a shortlist of the buyers' requirements. These buyers are not going to buy a product, whatever the price, if it has no utility; it is unlikely they will buy it if it does not correspond to their identity. As examples, companies like Volvo offer a value proposition based on safety and durability; Domino's Pizza base theirs on a hot pizza delivered to your door.

In the realm of project management, we like to calculate value as follows (Equation 2.1):

■ Equation 2.1: Value = Quality/Costs,

Interestingly, budget (or cost) and quality are two considerations that match the iron triangle used in project management, as shown in Figure 2.9.

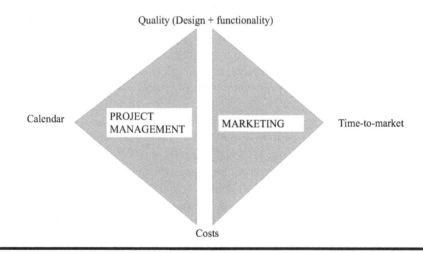

Figure 2.9 The iron triangles of marketing and project management.

Note: This diagram shows the intricate relationship between marketing and project theory, with both aiming for quality and reasonable costs. They both also aim for what marketers call "time-to-market": the speed at which innovative products enter the market.

There are three forms of value, as previously mentioned. **Perceived value** is in the eyes of the customer. It is often based on a comparison between benefits and cost (more exactly, sacrifices), including such consideration as time spent to acquire the product and psychological cost (such as dealing with an aggressive salesperson). Some marketing books posit that perceived value increases according to what customers gain (benefits) and what they give (not give up). (We consider this an "arguable error.") Customers do not think in terms of giving, but rather in terms of giving up (in terms of lost opportunity). For example, they give up personal time by having to go buy train tickets. From our perspective, it is more accurate to say that perceived value increases as benefits increase and sacrifices (what is given up) diminish.

Some customers may be willing to pay an outrageous price for a diamond ring because, in their mind, this will communicate their love to their lover. This is perceived value. **Added value** is the value added at every step of the production process; it is the sum of the accounting costs once the product is finalized (value chain). Some people treat perceived value as added value, but this is an "arguable error." Astute marketers know how to play with these two kinds of values: They will add a low-cost attribute on a product (such as a free handbag) to an expensive product set (cosmetics), but actually charge a premium price for the entire package. **Residual value** is the value leftover once the product has gone through its useful life, as set by the customer who bought it. Toyota initially built its reputation on residual value since its used cars sold at a much higher price than their American counterparts so that, in the end, customers who bought Toyota cars enjoyed a better value compared to American cars.

Simple definition: **Perceived value** is in the eyes of the customer.

Added value is the value added at every step of the production process.

The logic in terms of loyalty is as follows: High value brings loyalty. High value means increased satisfaction, which translates into repeat purchases. This goes back to one of the core models we saw previously. Of note, the opposite of satisfaction is not dissatisfaction, contrary to what some marketing books claim, and many marketing questionnaires posit, we consider this an "arguable error." Hate is not the opposite of love; because you do not love someone does not mean you hate her/him. Similarly, because customers are unsatisfied does not mean they are wholly dissatisfied. Being dissatisfied implies a negative feeling and a likely absence of repeat purchases. Being unsatisfied means the product did not meet expectations, but does not automatically generate harsh feelings.

To build value, marketers need to understand their customers (**customer deciphering**, if we can use this term). What is it that appeals to their senses?

What functions or utilities do they look for in products? What price would entice them to get off their couch and go to a point of sale to buy the product?

The tools for loyalty building are loyalty programs, the inescapable promise, and customer deciphering (based on design, functionality, and costs).

CHAPTER 2, CLASS EXERCISE #6:

Think of a situation where a company has an established loyalty program, but has refused to honor it. Discuss it.

PROPOSED QUESTIONS FOR DEBATE:

1. Do companies use unethical means to guarantee loyalty (e.g., locked-in contracts)?
2. Should there be better laws to protect consumers against unfair contracts?

2.2.7 Summary

We summarize the six components of marketing strategies in the context of the marketing feasibility of projects and what we have learned so far, as follows (Table 2.7).

Table 2.7   A Summary of the Six Components of a Project-Related Marketing Plan, and Their Tools

Component	Summary of Tools	Examples of Concrete Applications
Innovation	Brainstorming, cross-field imbrications	Patent, authors' rights
Segmentation	Four pillars: socio-demographic, geographic, psychographic, and behavioral	Needs analysis
Positioning	Perceptual maps, Cartesian maps and tables (spreadsheets), multi-criteria analysis	Competition assessment
Targeting	4Ps: product, price, promotion, and place; culture management	Policies, ads
Differentiation	Perceptual and mnemonic tricks, apprehension model, brand management	Brand-promoting campaigns, clever product, and message conception
Loyalty	Loyalty programs, personality deciphering, promise	Exclusive clubs

Project managers who understand these six core components of a marketing strategy are well equipped to communicate with their counterparts in the marketing department. The marketers who understand how these marketing concepts correlate with project management theory are most likely to build mutually beneficial relationships with project managers.

2.2.8 The Marketing Plan

Marketing plans vary in length, content, and format depending on the ones doing them, the depth of the qualitative and/or quantitative research, the intended audience, and the budget allocated for their completion. Chapter 4 covers the marketing feasibility of projects and its associated planning process, while during our seminars we provide templates for preparing various marketing plans and feasibility analyses. Readers will adapt these tools to meet their own requirements, so long as they ensure that they cover all important aspects of the business that allow a marketing feasibility analyst to decide whether the project should go ahead (→), be revised (←), or else abandoned (↓).

CHAPTER 2, CLASS EXERCISE #7:

Prepare a simple marketing plan and present it to your group. A convincing presentation is of the essence.

PROPOSED QUESTIONS FOR DEBATE:

1. Should you alter data to fit your needs for the presentation?
2. Should consumers trust statements such as "90% of experts say this product is best?"

2.3 Conclusion

We dedicated the above discussion to a general introduction to marketing concepts of which project managers should be aware. Teachers or tutors can also use it to present marketing to students who have little knowledge about the matter. There is much more to marketing, but recall that a complete marketing plan certainly includes the six components of the marketing strategy. It covers how much of the **market share** the company wants to have (e.g., Coke and Pepsi each have approximately 33% of the world market for their products: sweetened soft drinks). A complete marketing plan also includes forecasts, discussing expected sales and revenues; investors want to see these numbers and will use them when analyzing the financial feasibility of projects.

CHAPTER 2, CLASS EXERCISE #8:

Choose a single product and its brand, and prepare a PowerPoint presentation reviewing each of the six components of the marketing strategy.

PROPOSED QUESTIONS FOR DEBATE:

1. Is a convincing PowerPoint presentation useful in attracting investors' attention?
2. What can we do when data is missing? (Clues: Use proxies, use estimates based on similar markets or products, etc.)

Normal marketing plans necessarily include a section on decisions the company must make to grow the business,[20] but for the sake of a feasibility analysis, this is out of focus. We recommend the reader researches the topic independently.[21]

Project promoters work carefully to convince potential investors that their projects, once operationalized, will be profitable (which does not mean feasible). A project could well be feasible, yet not proved profitable. Conversely, promoters may excitedly try to convince a bank their projects are profitable, only to find out that they are not yet feasible because the necessary infrastructures are not in place (for example).

Marketing feasibility of projects analysts determine whether the project is marketable. Is there a need, an opportunity? Is there a market? Do the promoters capture the real psychodynamics of the would-be end users (consumer deciphering)? More specifically, have they carefully considered

1. The attractiveness/utility of the proposed innovation?
2. The market segment of consumers who will find more benefits than sacrifices in the project?
3. The position of their offering against competitive or other offers across time?

---

[20] Typically, marketing plans have general strategic (e.g., defining the mission), functional (e.g., devising the six components of the marketing strategy), and operational (e.g., preparing the action steps) sub-plans.

[21] Concepts worth studying when wanting to actually implement a marketing plan—an activity outside the scope of this book—include market defensive strategies (e.g., divesting, flank, preemptive, contraction, mobile and withdrawal defenses, and counteroffensives) and offensive strategies (e.g., guerilla, bypass, frontal, leapfrogging into new technologies, flank, and encirclement attacks). Growth strategies include brand development and innovation, changing packaging, diversification, harvesting, modification or proposing news usages, upgrades or downgrades, developing alliances or new partnerships (cross-merchandizing or licensing), horizontal or vertical acquisitions, market expansion, and many others. This book focuses on innovation.

4. The target segments and appropriate product/price/promotion/place policies?
5. The development of a differentiated approach?
6. The planned-for long-term relationships with the community and the end users?

If it does not meet these criteria, regardless of the profit the promoters expect to make, feasibility experts deem the project not feasible (↓) from a marketing point of view.

One way of getting a quick, intuitive image of the feasibility of a project is by listing the movers, the blockers, and the delayers.

The **marketing movers** are characteristics that incite people to want to move ahead with the project as it looks promising. Most often, project promoters emphasize the marketing movers and minimize, or ignore, blockers and delayers because they are eager to access funding and/or because they do not want to face the reality of their projects—which, in a large percentage of cases, suffer from blockers and, at times, delayers. Marketing movers are proper innovation, segmentation, positioning, targeting, differentiation, and loyalty-building strategies.

The **marketing blockers** are characteristics that fail to motivate stakeholders (in particular, the investors) to move forward with the project; they invite inertia. Should the promoters manage to palliate these characteristics, funders would likely adopt the project. Blockers are often elements external to the project, such as in a project of building a school in a developing country where there is a lack of electrical infrastructure. The promoters can find a way to respond to this challenge, without amending the project substantially, for example by engaging the local government. Marketing blockers include lack of patents or legal protection, poor selling infrastructures, intense or illegal competition, diluted markets (consumers are spread; e.g., in a state of war), lack of competitive advantage in relation to the key attributes, and a volatile consumer base.

Marketing delayers are characteristics that pull the project backward (utility drawback); they represent a serious drawback promoters can hardly overcome unless they revise it considerably. Marketing delayers include an unrealistic timeline, unbearable costs, and an inability to apply, control, and enforce quality standards.

Table 2.8 illustrates these intuitive concepts.

Marketing feasibility experts can use Table 2.8 on an intuitive basis when promoters present their projects. For having heard many conversations between promoters and potential investors or else marketing experts, we feel confident this brief assessment provides a sound analytical overview. Often, the mere fact of not finding enough movers, and of facing considerable amounts of blockers and delayers, is enough to encourage promoters to revisit their projects and remodel their ambitions. Experienced marketing professionals will promptly recognize if a project suffers from, say, an inadequate level of innovation or else targets an overly volatile market.

Table 2.8 Intuitive Marketing Movers, Blockers, and Delayers

Movers	Blockers	Delayers
Proper innovation	Lack of patents or legal protection	Unrealistic timeline
Proper segmentation	Poor selling infrastructures	Excessive costs
Proper positioning	Intense or illegal competition	Inability to apply, control, and enforce quality standards
Proper targeting	Diluted markets	-
Proper differentiation	Lack of competitive advantage in relation to the key attributes	-
Proper loyalty-building strategies	A volatile consumer base	-

Rule of Thumb: In marketing projects, seek movers and avoid blockers and delayers.

To assess the marketing feasibility of projects, experts must look at each of the six components of marketing strategy, which includes needs/opportunity assessments, and determine if they are convincing and well supported by well-documented hard data. If it is not the case, the project promoters are not going to be able to market their projects, which means they will most likely not be profitable. Again, a marketing plan often includes assumptions difficult for the marketing feasibility expert to assess, forecasted sales being an example. The best way to determine whether the product can meet the sales forecast is to think upstream, and to do that we need to assess the marketing feasibility of projects. To do so, experts carefully examine the form and content of the six components of the marketing strategies as presented by the project promoters. They leave the forecasts and revenue expectations to the financial feasibility analysis because marketing feasibility analysis does not try to determine if a product will earn money, but only if all the ingredients are present to make the product's marketing plan feasible. This distinction is fundamental. Certainly, marketers can work out the numbers and team up with the financial experts but, ultimately, marketing feasibility analysts only look as to whether the project and its deliverable are feasible (not profitable).

Again, marketing feasibility experts are not interested in empty promises and in forecasting sales enthusiastic promoters present to sway investors in their favor. Their job is merely to assess if the six components of the marketing strategy present precarious assumptions and data, or else, if they are solid and anchored in sound, objective judgment.

We did not discuss international marketing to any great extent as managers must decide if they want to go abroad (e.g., export a project, such as a rock concert), which markets to enter, how to enter them, and then work out a marketing plan. (This is reserved for another book.)

2.4 Mind Teasers

Readers may use the mind teasers as questions in preparation for an examination or quiz.

1. Briefly describe
   a. the difference between length and depth of a distribution chain and
   b. the six criteria for proper segmentation.
2. Briefly explain
   a. the four pillars of segmentation and
   b. the notion of differentiation bilateralism.
3. Define
   a. backward engineering,
   b. brand equity,
   c. CLV, and
   d. the initial value proposition of a project.
4. Differentiate between
   a. skimming and market penetration and
   b. primary and secondary data.
5. Draw a
   a. standard product's lifecycle (with all necessary information),
   b. two-by-two table explaining entry strategies, positioning projects in the appropriate quadrants, and
   c. typical communication model between a seller and a buyer.
6. Give
   a. a mathematical rendition of the notion of value,
   b. an example of multi-criteria analysis,
   c. at least four qualities that reinforce a brand's equity,
   d. at least four steps experts have proposed to ensure consumers/end users adopt a new product,
   e. at least four things marketers do to generate symbolic value,
   f. at least three pricing strategies,
   g. at least two modes of communication marketers use to relay their messages,
   h. some examples of ads built on differentiation,
   i. the elements contained in any marketing mix,

j. the three criteria innovative products must meet,

k. three common causes of product launch failure, and

l. two types of segmentation, besides macro-segmentation.

7. List, and briefly define, the six components of marketing plans for marketing feasibility of project analysis.

8. List and briefly describe

a. the 4Ps of marketing,

b. at least four assumptions marketers make with regard to market segments,

c. at least four benefits brands offer, and

d. at least four elements innovations inspired by nature generally display.

9. True or False?

a. Perceptual maps are based on products' attributes.

b. With respect to customers, marketers often focus on their readiness to buy; their decision processes when alone, in pairs/couples, or in groups; the benefits they seek; the usage they make of the products, whether regular or inventive; and their levels of infidelity.

c. A great adequacy occurs when the attributes of the product match the attributes of the designated market segment (or target market).

d. A promise is when the regulators make buyers believe they are purchasing the right product, and should hence become (or remain) loyal.

e. Added value is the value included at the end of the production process.

f. All filters affect the flow of communication and are, in essence, anti-cultural.

g. Differentiation is about ensuring consumers recognize our products, which sit on a shelf that harbors hundreds of similar products.

h. Marketing plans vary in length, content, and format depending on the author, the depth of the qualitative and/or quantitative research, the intended audience, and allocated budget.

i. Perceived value is in the eyes of the consumer.

j. Points-of-parity (or resemblance) and points-of-difference are not related to the notion of competitive advantage.

k. Positioning is to products what segmentation is to people.

l. Primary attributes are things customers absolutely want, and not having them will discourage these would-be clients from buying the product.

m. Promotion and publicity are one and the same.

n. Targeting uses three out of the 4Ps of marketing.

o. The third concept of importance, with respect to price, is that of placement.

p. To promote loyalty, marketers want to ensure consumers feel safe and secure (with little or no perceived threat), and feel they can trust the product because it delivers what it is promised.

q. When marketers are engaged in the decision-making process, they do not consider different scenarios.

    r. Segmentation is about products, and products only.

    s. Product creators balance functionality (including compatibility) and design (including ease of use).

10. What is/are
    a. the role of opinion leaders;
    b. a market share;
    c. market movers, blockers, and delayers;
    d. filters in a communication model?

Chapter 3

What Is a Project?

The staircase in City Hall, in Nancy, France. The Art Nouveau décor was inspired by nature, with particular botanical and insects' shapes. Nature often inspires innovation.

3.1 Introduction

Before starting a project feasibility study, we must first understand what projects are and why they fail. The number one cause of failure stems from project promoters not properly defining the project from the outset. The proper definition of a project is "something" that is specific, measurable, attainable, relevant, and time-bound (SMART), nothing less. We define that something in this chapter.

The word "project" derives from the Latin words *projectum* and *prociere*, which translate into "to throw something forward."[1] Hence, all projects involve some form of anticipation.

> Simple definition: **Projects** refer to undertakings that prepare future operations.

Experts developed terminology for project management in the 1930s,[2] and first used the term "project manager" in the *Harvard Business Review*.[3] It was only in the second half of the 20th century, however, that project management as we currently know it developed,[4] with its theory growing more complex over the years.[5] More precisely, rebuilding efforts following the end of World War II generated numerous large-scale projects in both Europe and Asia.[6] By the early 1950s, the U.S. defense and aerospace sectors started using new techniques and terminology linked to project management[7] and companies worldwide henceforth realized the value of this developing science, including project portfolios and certification programs.

Project Management Body of Knowledge (PMBOK) considers projects as composed of ten management areas, and each implies some managerial action and/or specific job title:

1. Calendar,
2. Costs,
3. Quality,
4. Resources,
5. Communication (which we relate to marketing),
6. Risks,
7. Supply,
8. Stakeholders,
9. Project integration, and
10. Work parameters.

In reality, managers rarely organize their projects in full accordance with this list. In fact, resources and supply are often combined, and communication

---

[1] Dalal (2012, p. 2).
[2] Morris (1994, 2011).
[3] Gaddis (1959).
[4] Hughes (2013).
[5] Baccarini (1996).
[6] Shi et al. (2015).
[7] Morris (2013).

and stakeholders often fall within a marketing or pseudo-marketing group. PMBOK is, therefore, more of a general guideline than a template managers fully replicate.

The learning objectives of this chapter are to understand what projects are, to recognize them in daily life, and to develop a sense about how to treat them from managerial and marketing perspectives.

3.2 More on the Definition of Projects

Experts conceive of projects in different ways. For some, projects are made up of benefits, hypotheses, and plans, while others define projects:

- As any undertaking that may benefit from project-management theories, whether small (10,000 to 100,000 engineering hours), medium (100,001 to 500,000), or large (over 500,000);
- As sets of "interrelated units" that have a common purpose[8];
- As systems composed of various personalities, manpower, technical expertise, priorities, schedules, and cost challenges[9];
- As temporary efforts for creating products or services that have a beginning and an end, and can be spread over the short, medium, or long term[10];
- As undertakings that aim to improve people's living conditions (e.g., vaccination programs), processes (e.g., new information systems), or technical devices (e.g., power-based structures such as nuclear plants);
- As ventures that comprise inputs, a transformation phase, and outputs;
- As ventures that create value for promoters, investors, and end users; and
- By their programs, goals, and impacts.[11]

For clarity, we define a project as

> ... a concrete and organized effort that leads to the realization of a unique and innovative deliverable, which can be a product, service, or process, or even a science research initiative, which is conceived based on a perceived opportunity. The project has a beginning and an end, which can sometimes serve as the new bedrock for a different project. It involves a plan, some processes, people, and a line of authority; it contains inherent challenges and problems. It is bound by a specific calendar, a cost structure, and preset norms of quality. Finally, each

---

[8] O'Shaugnessy (1992, p. 10).
[9] Morris & Pinto (2004, p. 9).
[10] PMBOK, fifth and sixth editions.
[11] See Dingle (1985).

Table 3.1 The Eight Characteristics, and the Three Functions, That Define a Project

Eight Structural Characteristics
1. Responds to an opportunity
2. Benefits from stakeholder involvement
3. Is unique
4. Is innovative
5. Is subject to quality constraints
6. Requires some investment
7. Contains challenges
8. Has a specific timeline
Three Functional Characteristics
1. Generates deliverables
2. Creates formalized knowledge
3. Has some impact, positive and/or negative

project tends to generate some official documentation as well as positive and potentially negative impacts.[12]

Projects must meet the eight characteristics mentioned in Table 3.1. If they do not meet these criteria, they are not projects in the pure sense; they are operations, ideas, or fantasies—anything but a project.

Uniqueness and innovation are not the same concepts. For example, a project may be unique in that it involves building a casino in an area that does not have another one within 100 square miles, but that does not make it innovative. To be innovative, a project must meet the conditions as described in Chapter 2 under Innovation.

In essence, what the promoter presents is a project in the purest sense, that is, in the sense that it requires a feasibility study and not, for example, the repetition of something already created (in which case it is not unique, and hence has proven feasible). One must ensure all eight *sine qua non* structural characteristics, and at least two of the three functions, are present in the projects they conceive. That is the rule!

---

[12] Mesly (2017).

CHAPTER 3, CLASS EXERCISE #1:

List and describe some projects in which you have been involved.

PROPOSED QUESTIONS FOR DEBATE:

1. Are all projects ethically sound?
2. Should governments invest in partnerships with the private sector?

We represent projects in Figure 3.1.

A project's time boundary is set at the beginning, the input, and at the end, the output, acting as walls of sorts. Cost constraints act as a ceiling (maximum level to be reached), and norms of quality act as floors (minimum level to build the project on). Typically, we segregate inputs into two sections: one mechanical and the other human. The mechanical section includes plans and processes. Plans are what promoters prepare when explaining projects to potential investors, stating their intended future actions to achieve their goal. Processes are all the repetitive mechanisms that ensure the project progresses; for example, hiring staff involves a process of placing an ad, collecting curricula vitae (CVs), and interviewing candidates. Starting a corrugated cardboard machine and feeding the material through is a process that contains a preset number of specific steps, from which the operator cannot deviate.

The second section, the human one, consists of people and power. "People" refers to the general staff, those who do not make strategic decisions for a project. We classify such staff into four categories:

1. The internal and external producers (forces of production),
2. The consumers and end users, and other such stakeholders;

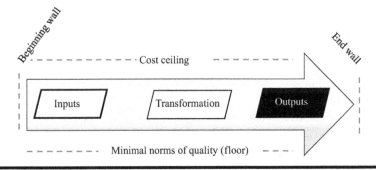

Figure 3.1   Projects.

Note: This view of a project is original, yet one that allows anyone to understand quickly what projects are all about.

3. The internal and external regulators, like accountants or government; and
4. The "bad apples": those who interfere with the project (e.g., disruptive employees or independent groups seeking to disrupt the project).

"Power," on the other hand, refers to the line of authority in terms of structure; any individual who makes strategic decisions with respect to the project is part of power. Project management typically segregates between horizontal, vertical, and **matrix** (a mix of horizontal and vertical managerial structures) organizational charts.

In regard to outputs, we categorize them into two sections: **tangibles** and **intangibles**. Tangibles include three components: the deliverables, the knowledge acquired, and the impacts, which can be positive and/or negative. Intangibles are phenomena that promoters may want but do not and cannot control or forecast, such as group bonding and pride (two social phenomena that appear regularly within projects).

As already briefly discussed, project deliverables come in four forms: products, services, processes, and research. Processes refer to such undertakings as improving the flow of consumers in a store or in a train station by way of redesigning the floor plan, putting "obstacles" to prevent people from taking certain paths, or placing better signage to direct traffic more effectively. Indeed, in this case, the core of the project is one of the processes: improving a flow. Research talks more particularly to the academic community; normally, funded projects imply a cost ceiling and include deadlines by which researchers must prove to investors (often governments) that they achieved their goals. PMBOK expands processes by dividing them into five categories, each potentially headed by a manager: initialization, planning, execution, management, and closing. This, of course, espouses a project's lifecycle, which goes from vision to planning, mobilizing, deployment, and completion/measurement.

Knowledge is also at the core of all projects. Because, by definition, each project is unique (and innovative), project managers acquire new skills and information about people and processes, in particular. For large projects, managers usually create some book or booklet of knowledge (named differently, depending on the author) that serves as a measurement of what managers have accomplished; it also serves as a tool for future, similar projects. We separate formalized knowledge into four categories, with documents related to

1. Plan (the initial blueprints upon which the project was undertaken; the project charter),
2. Processes (the day-to-day and well-defined transformation steps),
3. People (the stakeholders, including employees, funders, and officials), and
4. Power (the line of authority).

Plan-related documents include general ones such as risk models, and the project charters. Process-related documents encompass quality control and performance measures (e.g., Material Safety Data Sheet—MSDS). People-related documents comprise, of course, CVs and work contracts. Finally, power-related documents relate to organizational actions and charts (organigrams), including financial reports and governance. As seen in Chapter 2, marketing experts will want to learn from these aspects of project management because the two fields share many aspects related to business. There is no marketing strategy worth that name that does not take into due consideration planning, processes, people, and power. Marketing includes leveraging the power of consumers (end users), which they may do through the Internet, social movements, promoting an individual's constructive behaviors; and eliminating destructive behaviors (such as stealing).

As for impacts, they are simply inevitable. They are effects that result from a project, yet they are not a part of the project itself (e.g., collateral damages or benefits). Examples include noise or dust pollution during the construction of a bridge, or bringing additional visibility for a city (e.g., New York after finishing the One World Trade Center). Promoters who pretend their project will have no impact (especially no negative ones), do not understand their projects, and do not deserve funding. Marketers should be equally conscious of the events or behaviors they may unleash into the marketplace through their actions; only then can they design the best, and most productive, actions for the project under consideration.

3.3 What Is the Iron (Bermuda) Triangle?

We have seen the **iron triangle**[13] from the onset of this book. The iron triangle is simply the calendar of tasks and activities, the budget, and the preset standards of quality. All these are measurable. Books that pretend the iron triangle consisting of timeline, budget (or costs), and scope make an "arguable error" (and we would say a huge one). Scope is not measurable. At the end of a project, the experts can justifiably claim that the project was finished on, say, September 1, 2015 as promised, that it did not exceed the budget of C$ 400 million, and that the deliverable passed all quality tests required by government regulations and engineers. The experts cannot put a similar scale on scope: Scope is, simply, not realistically measurable. Satisfaction is not measurable in an objective way either. Projects are not about satisfaction; books that pretend this is the case are making an "arguable error" that may misguide their readers. Also, books that say that the iron triangle is composed of delay, costs, and whatever (scope or norms of quality) also make an "arguable error." Delay is actually the last thing the funders want in a project, besides cost overruns and quality issues that will haunt them for years to come. Recall that in feasibility analysis, we are only concerned with what is measurable, nothing less, and certainly nothing that is left to personal judgment. In the exercises that we provide during our seminars (see Appendix 1), we pay much attention to training

---

[13] We like to use the term "Bermuda triangle" to outline the fact that its three components can lead to disastrous effects when not well managed.

the would-be marketing feasibility analysts to recognize and deal with personal judgments and biases. These are hard to admit and leave many participants with the feeling of emptiness or self-defeat, but the job is to be as objective as possible, not to rely on personal opinions. Inasmuch as it is a difficult task, it is worth it: Well-evaluated projects have that more chances of succeeding. By being objectively critical, we, as marketing feasibility analysts, do our clients and/or the promoters a huge favor: We save them time, money, efforts, and possible embarrassment.

The challenge with the iron triangle is that it is very hard to conciliate the demands of calendar, costs, and norms of quality: They tend to antagonize each other, like magnets repulse each other when their polarized sides face each other. This is why many projects fail and why feasibility analyses are needed, including and necessarily marketing ones.

As discussed in Chapters 1 and 2, marketers make a **promise** to their customers: The product will do what it has been designed to do. Project managers make a promise to the funders and end users: We will complete the project on time, within budget, and in full respect of the preset norms of quality. This, in turn, will ensure satisfaction and pride. Marketers phrase their promise using a language that appeals to their target audience, ideally based on the three primary attributes that condition buying. Project managers simply use the iron triangle as the base of their promise. A project, by definition, is a promise. Hence, every step of the way, project managers are marketers; marketers, by preparing their plans with respect to the iron triangle, act as project managers.

CHAPTER 3, CLASS EXERCISE #2:

Pick a project in which you were involved, even a personal one (e.g., planning and going on vacation). Examine it in terms of the iron triangle and its underlying promise.

PROPOSED QUESTIONS FOR DEBATE:
1. Should all projects hold a promise?
2. Based on your own experience, why is it difficult to balance the three aspects of the iron triangle?

3.4 What Are a Project's Lifecycle and Key Components?

No project is static. Rather, projects are dynamic entities that follow a typical pattern from concept through to their realization. Project managers refer to this as a project's **lifecycle**; in fact, such cycle shares many similarities with a product's lifecycle, which we saw previously. The curves of both lifecycles are identical, and for both, time is of the essence, of course.

We separate a project's lifecycle into five phases, plotted by time (or, more precisely, calendar of tasks and activities) on the X-axis and efforts on the Y-axis. We can quantify efforts by attributing them a monetary value, as shown in Figure 3.2.

These five phases deserve an explanation as they are labeled differently depending on the book referenced. The first phase involves vision: having a dream. Nothing is decided, but the project holder (promoter) has some idea as to a way to fulfill a need or respond to a market opportunity. A smart thing to do at this point is to share the idea with marketing experts, as they have sound knowledge and a feeling for the pulse of the market. This is, in large part, where project feasibility—the focus of this book, of course—begins.

The second phase is planning. Again, marketers have a role to play here as planning is about preparing to answer the needs of two groups of stakeholders. The first are the investors, who want to see a solid proposal covering all relevant aspects of the business (including marketing, financial, and technical). Second are the end users, those who will benefit from what the deliverable (being produced by the project) has to offer. Planning for large-scale projects necessarily involves feasibility (and often prefeasibility) studies. From a marketing point of view, this means determining whether a project will find acceptance with both groups of stakeholders. Investors are concerned with financial, and possibly social, needs: receiving a return on money invested and/or providing end users (such as a community) with deliverables that improve their lives. The end users (or clients) want their

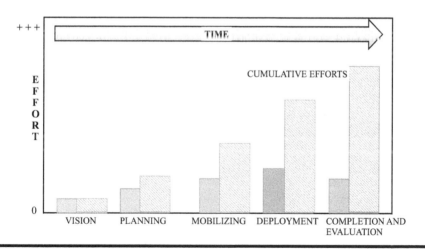

Figure 3.2 Efforts and phases.

Note: Each project phase involves a certain level of effort. These efforts accumulate over time, so it gets harder to abandon a project even when its chance of success appears slim.

expectations met, whether relating to essential accommodations, such as access to safe water, or entertainment. They seek satisfaction.

The third phase refers to mobilizing all raw materials and energy sources (often provided by community infrastructures such as electrical stations), as well as the **forces and the means of production** (whether onsite or at remote locations), with the objective of having them readily available. Client participation in projects is normally minimal at this phase, which is purely operational. Forces of production are people; for example, a group of autoworkers lined up along a production line. Means of production are the robots and tools they use to perform their assigned tasks.

The fourth phase is deployment; this occurs when project managers activate plans and put into production the raw materials, the energy sources (often, regretfully, ignored in project management books, which we find is an "arguable error"), and the forces and the means of production mobilized in the prior phase. Clients who participate in the projects may have a very active role at this point, one where disagreements with project managers may occur. Clients may change their minds as to what they want, or else project managers may realize they cannot accomplish what they had promised to do.

The fifth and final phase is when project managers submit deliverables to the final end user, whether a government, a private enterprise, a group, or an individual. Past this point, the deliverable is no longer a project, but rather an operation. Project managers, investors, and would-be end users usually celebrate the achievement to which they have contributed in their own way from the vision phase onward. This last phase necessarily includes measuring the project, that is, going back to the calendar, budget and norms of quality in an informed and well-documented way. Books on project management that ignore the fact that all projects must be measured in such a way at the end make an "arguable error." How, as a funder, will you ever assess if the feasibility analysis for which you paid a high price, was worth your while if you only rely on a few customers' happy faces or a political speech priding the efforts of everyone? This cannot be serious. Projects deserve a better outcome, and certainly, future projects will benefit from the knowledge dutifully recorded from your project. To illustrate this, think of projects as carving the paste of history. Your footprint will stay for generations to come; it had better be outstanding and proven to be as such.

Hence, managers must enrich the completion phase with a full valuation of the project, whereby everything the stakeholders have accomplished and learned is examined, reviewed, and recorded in a book of knowledge (BOK). Again, they should fully and accurately measure the initial plan. This includes delivery terms, costs, and norms of quality. If anything, astute marketers will use the measures done on the iron triangle at the end of the project to promote it even more and better than initially thought.

PMBOK expands on the lifecycle by citing, and describing in detail, steps managers need to take to complete a project. They are as follows:

1. Preparing the **project charter**, based in part on such documents as the business case conceived to justify the project and written agreements and contracts;
2. Preparing the project's management plan, which in part includes scheduling meetings;
3. Managing the workflow, which implies creating the deliverable and keeping a journal of activities;
4. Managing the acquisition of knowledge, which means keeping a register of what has been learned and producing, in the end, the BOK (or some form of it);
5. Controlling the work and the workflow, by way of measuring performance;
6. Managing change (something we will discuss in project feasibility by way of contingency planning); and
7. Closing the project or a project's phase.

The above list, of course, belongs more specifically to the actual management of a project rather than to its marketing feasibility analysis. Certainly, however, a draft of a project charter, submitted along with other documents by the project's promoters, always helps convince potential investors of the seriousness of the proposal.

3.5 What Are the 4Ps of Projects?

Since the 1960s, marketing experts have relied on the 4Ps of marketing: product, price, promotion, and place (distribution). Project managers have a 4P system of their own, which we have touched on many times so far: plan, processes, people, and power.[14] Managers use the 4Ps of project management to drive projects in a positive way. Once engaged in the production process, we label the 4Ps with an apostrophe, as follows: plan', processes', people', and power'. This is a simple way of saying managers are using them proactively, meaning the 4Ps will now evolve with time. Figure 3.3 illustrates how project managers use the 4Ps.

As mentioned, plans and processes constitute the internal, mechanical elements of a project. People and power deal with the psychodynamics and organization of a project, something often dealt with rather lightly in many project management books, despite their importance. To treat the human aspect of projects with little regard is an "arguable error." As we will proceed to show in Chapters 4, 5, and 6, most projects fail because of their points of vulnerability (POVs), and all POVs relate to the human factor. The PMBOK uses the terminology of people in its own way, with the "execution group" corresponding to people, and the "control group" to line of authority (power). It, however, ignores the role of clients who participate

---

[14] In French, "Plan, Processus, Personnes, et Pouvoir"; and in Spanish, "Propósito, Procesos, Personas, and Poder."

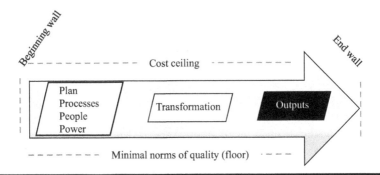

Figure 3.3   Projects and the 4Ps.

Note: Projects have 4Ps of their own, which are a useful way of comprehending projects. Missing one inevitably means the project is not adequately designed.

in the project and/or in product development. The 4Ps are *sine qua non* considerations the project feasibility analyst must take into consideration.

We have spoken a few times about **structural variables**. To recognize such *sine qua non* conditions (for the 4Ps or for the structural characteristics seen below), one must ask the following two questions:

1. Could be project without a plan (or without processes, people, or power)? and
2. Can each of the 4Ps be set independently of one another (before being used in the transformation phase)?

The answer to the first is "no," and confirms that a project, without a plan (processes, people, power), does not exist. The answer to the second question is "yes" and means that, statistically, before the transformation phase, each of the 4P bears no correlation with the other Ps. Indeed, at the beginning of the project, prior to engaging and using resources or energy, managers may treat each of the 4Ps independently. Hence, the 4Ps, at this point, are simply structural variables; there is no firm correlation between each. Think of a bicycle and its various parts. The seat, chain, and two wheels exist independently from one another; however, to make a true bicycle, we need each of these components. Each element is a *sine qua non* condition to building the bicycle. The same logic applies to the 4Ps in above question (1): Each P is mandatory.

Prior to the transformation phase, the answer to question (2) is "yes." Project managers can set up a plan without having determined whom they are going to employ, or who will be the floor manager. Each P is independent of one another (in the pre-transformation phase). Indeed, the 4Ps will only combine and interrelate during the transformation phase. For now, the promoters (power) can decide the plan independently of the people and of the processes. We consider the 4Ps as inputs, each with its own weight on the outcome of the project.

A substantial change occurs during transformation, one that affects project and marketing managers, as well as clients: For example, whatever people do affects the line of authority (power), and vice versa. A miscarriage of a process will affect the plan. All the 4Ps combine, and things become a whole lot more complex and dynamic: POVs surface, such as when clients request changes, project managers did not expect. The 4Ps are no longer independent from one another.

Based on our experience and research, we estimate POVs are at a minimum of 4%–7%, which is significant. Many of these POVs concern both the project and marketing managers, which highlights the importance of their strong cohesion and mutual understanding.

Of paramount importance to marketing and project managers is defining the 4Ps during the preliminary analysis. Of course, managers can hardly anticipate all interactions that will take place once the project has commenced, yet a proper assessment ensures mutual understanding and helps avoid POVs.

Rule of Thumb: The higher the number of POVs and the weaker the remedial actions are, the less the project is feasible.

Project management theory divides groups according to certain functions. The execution group, as the name indicates, belongs to people: They obey orders and produce product. The surveillance group belongs to power, because it implies some form of authority and decision-making involving others; at minimum, it includes project managers and some key staff. The control group, for obvious reasons and as mentioned, also belongs to power. Generally, the feasibility analyst can determine these by listing what these individuals do: order, command, direct, impose, control, and so on. Within the control group, project leaders plan for the long term and share their vision. Project managers are task-oriented; they administer instructions, act to minimize risks and vulnerabilities, drive project activities, set and impose or supervise costs parameters and schedule, lead and motivate teams, and foster a certain culture within the members of a project.

The marketing expert must understand the role of people, acknowledging their competencies and recognizing that a positive 4Ps culture, leadership, and eagerness to perform provide an additional boost to productivity.[15]

Some authors segregate people and power according to various criteria, such as customers (people), project teams, and parent organizations (power).[16]

---

[15] Mantel et al. (2011).
[16] Meredith and Mantel (2009).

The PMBOK[17] does not refer to people who actually cause havoc, voluntarily or not, during a project ("bad apples," as referred to before), regretfully. Yet, as we will see in Chapter 5 on the relationship between marketing and project managers, clients who participate in projects may become quite disruptive at times. An example is that of the Montréal Olympic Stadium, planned for the 1976 Olympics, which was never completed. The Commission Cliche inquiry found that various groups illegally benefited from it and caused major delays.

We can also divide people into two broad categories: external (such as banks or other similar stakeholders, promoters, or lenders) and internal people. Because external people are, by nature, outside the organization, we correlate them with risk. Indeed, some of them may endanger projects: environmental groups who may interfere with development, clients who do not participate in the actual realization and may cause frictions, or investors who may walk out. Internal people are associated with POVs, as all POVs ultimately relate to internal people. Machine dysfunction is not the machine's fault: People build and operate the machine.

CHAPTER 3, CLASS EXERCISE #3:

Choose a concert you enjoyed very much, whether classical or pop rock! Try to list rather randomly elements of the 4Ps. Thereafter, classify your findings according to these 4Ps.

PROPOSED QUESTIONS FOR DEBATE:

1. Is one P more important than the others before the project starts (at the vision phase)?
2. Could one P be more important than the others (e.g., processes), depending on how advanced the project?

3.6 What Are a Project's Points?

A certain number of "points" characterized projects. We will examine the POVs in more depth in Chapter 4. For now, we focus on two points that involve critical decisions that affect them in their own ways: the point-of-no-return and the point of autonomy. We mention them because they are often forgotten in project management books; this omission is an "arguable error." Even if project managers are unable to calculate these points in detail, for whatever valid reason, they must still be cognizant of them as they are crucial moments in the transformation phase—ones that affect the schedule, the costs, and the norms of quality tremendously.

---

[17] For example, see PMBOK 6 (2017).

3.6.1 Points-of-No-Return

When an avalanche starts, there is no stopping it, same as a tsunami. Similarly, in a project, situations evolve whereby one cannot go back in time; managers cannot reverse these evolving situations. The explosion of the Space Shuttle Columbia provides a painful example. While the engineers at the control center soon concluded the shuttle had suffered substantial damage during the takeoff, it was too late to call it back. In fact, calling it back would have made matters worse; the ground crew could only pray the problem would not be as dramatic as anticipated. Unfortunately, the damages sustained by flying debris included a punctured wing. After days in space, the shuttle disintegrated upon reentry into the atmosphere resulting from the extreme heat generated by the friction of air on the damaged wing.

All **points-of-no-return** come with particular costs. Stopping the process means all preceding processes and investments are now a net loss or, in terms of project management, a net utility drawback. Each of the constraints suffers: Managers have lost time, have had to reluctantly accept rising costs, and have incurred issues in quality.

From a marketing point of view, the point-of-no-return means that experts must implement drastic changes in preset marketing strategies should production fail to meet expectations, and this may mean launching a damage-control media campaign to advise end users of product (deliverable) changes. The goal would then not be to promote the project, but to minimize the negative impact of a problematic point-of-no-return. Again, in the case of the Space Shuttle Columbia, early on engineers strongly suspected, after examining videos of the launch, that the shuttle was severely damaged when a large piece of insulation foam broke off the external tank and struck its left wing at its most vulnerable point: one where extreme heat would build up upon reentry into the atmosphere. Yet NASA decided not to inform the crew or the public, betting the problem would not lead to catastrophe and that informing stakeholders would lead to a worse problem than the one faced by the damage.

3.6.2 Points of Autonomy

Project staff reach the point of autonomy when they can theoretically operate the entire transformation process, or the parts that concern them (e.g., a specific machine or process) without supervision. They feel competent about the machines they operate, or processes they handle, and know their limits so they can operate with their "eyes closed," so to speak. They have acquired all the necessary knowledge and expertise; they have fully integrated their learning curve.

When introducing staff to new machinery or processes, the number of person-hours is naturally high. Specialized trainers, consultants, or experts may accompany them in learning the basics of the tasks, and test the machines or processes extensively to ensure they can perform according to expectations.

There comes a point, however, when staff no longer needs help and knows the machines or processes inside out. This is the **point of autonomy**. Understanding when this occurs and what conditions are necessary to reach this point (hopefully, as soon as possible) helps both marketing and project managers plan other activities, because the time, money, and efforts they save can be efficiently dedicated to other tasks and duties. Typically, it is also a point where the clients who participate in the realization of the project feel fully confident in the capacities of the crew.

3.6.3 Other Critical Points: Benchmarks, Stage Gates, and Milestones

There are numerous other critical points project managers refer to in order to better structure their efforts and work with marketing experts to promote projects in their most advantageous light. Project managers know them well, so we only cover them briefly here. Figure 3.4 plots these points along a project's lifecycle.

All these points express critical moments marketing and project managers, and any extensive feasibility study, must take into account. They mark positive events within the flow of tasks and activities.

Benchmarks, represented by squares (as they are parameters), are closely linked with norms of quality. Managers will set levels that deliverables should meet, or the timeline it must respect, before the maximum costs are incurred. Typically, benchmark points are set according to predetermined standards established by

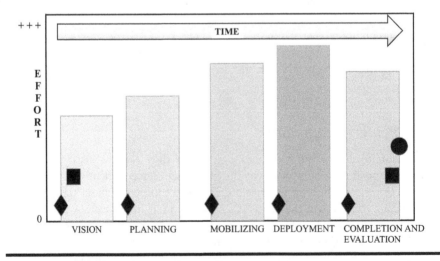

Figure 3.4 Points along the project's lifecycle.

Note: This figure positions various types of points along the project timeline and is often useful when working with customers/future end users.

organizations, by government guidelines and regulations, or from studying similar past projects.

Stage gates (represented by diamonds) mark the entry into a new stage/phase or substage/phase. At minimum, there are five stage gates, as the traditional model of project management comprises five phases, namely, vision, planning, mobilizing, deployment, and realization/evaluation.

Finally, **milestones** (represented by black dots) mark the achievement of an important step in the making of a deliverable (e.g., its delivery). They are sometimes celebrated publicly; certainly, such events help cement bonds between marketing and project managers.

CHAPTER 3, CLASS EXERCISE #4:

Choose a well-known project that was widely publicized, and identify some of its project points (such as milestones).

PROPOSED QUESTIONS FOR DEBATE:
1. Should marketers be involved in each project points (e.g., the point-of-no-return)?
2. Are advanced countries better at identifying and managing project points compared to less-advanced countries?

3.7 What Are the Key Consensus Factors (KCFs)?

Together, marketing and project managers form a group. For decades now, researchers have identified three *sine qua non* conditions that characterize groups within projects. They are as follows:

1. A common goal,
2. Some level of interdependence, and
3. Emotional relationships.[18]

These three elements glue teams together; they encompass, or else translate into, various concepts covered by academic literature, such as homogeneity and external risk.[19]

---

[18] Bales (1950), Bass (1960), Moreno (1969).

[19] Indeed, risk can sometimes reinforce cohesiveness in teams, as people need to join forces to face adversity. People adopt perceptions that help them form a coherent vision (Snyder and Uranowitz, 1978).

Recently, researchers have elaborated classifications in an effort to identify what promotes consensus among stakeholders; they list them as **key consensus factors** (KCFs).[20] These are by no means equal to trust or to cooperation *per se*; rather, they are often upstream of the group effort. Of course, consensus in turn fosters mutual trust and cooperation. In fact, we can define consensus as a group state whereby the members have reached a unanimous decision regarding a problem through trust and cooperation. KCFs are not equivalent to key success factors (KSFs) or key failure factors (KFFs).[21] The literature on project management seldom discusses KCFs (which we believe is an "arguable error"), although they are certainly instrumental in a project's performance.[22] Prefeasibility studies do not address KCFs, but feasibility studies certainly do (see Chapter 4 for more on this).

Under the current model, the KCFs amount to four, as follows:

1. A clear, common, and mutual objective;
2. A harmonized operational culture,[23] meaning shared backgrounds and experiences;
3. Mutual listening; and
4. A concern for meticulous work.

Combined, these KCFs promote the creation of value; they give a sense to the effort and bind stakeholders together. They also rely on and encourage trust, cooperation, and commitment. The model assumes they are universal; indeed, it is safe to assume people of all backgrounds and cultures are more likely to bond the more they share a clear and common goal (thus avoiding the dispersion of interests). When they have affinities based on past experiences and accumulated knowledge, they naturally listen more intently to each other and respectfully appreciate one another's point of view, and they pay attention to what they do. This improves quality while theoretically minimizing injuries and downtime.

On the field, project managers assume, rightfully or not, that staff will happily work together and agree on such things as information transfer and measures of

---

[20] Mesly and Braun (2019).

[21] In short, KSFs vary according to the type of organization (e.g., not-for-profit *versus* for-profit). According to the large majority of writings on the subject, they generally include a clear goal: well-defined impacts; access to necessary human and material resources; sound quality control; a realistic calendar; a well-balanced budget that plans for the unexpected; proven working methods and processes; reliable infrastructures; human relationships based on trust, cooperation, and commitment; support from top management; as well as a proper mix of transparency, fairness, and control. As for KFFs, they are vastly different from KSFs and not their mere opposite (it is an "arguable error" to think so): a lethal combination of a bad plan and an inadequate workforce—called the dreadful combination—diverging interests among stakeholders, and overconfidence (see Chapter 4).

[22] There are discussions in the general literature of the culture of consensus in Japan; for example, see Ouchi (1979).

[23] Recall we define culture according to the characteristics of the 4Ps of project management.

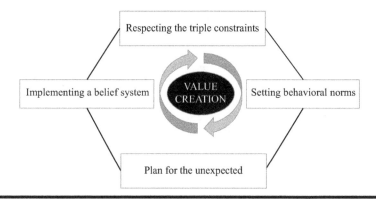

Figure 3.5   Stylized consensus and value model.

Note: Recall that projects bring value by responding to an opportunity. Customers build expectations toward the future deliverable and the capacity of the project's team to bring it to market. They long for value creation.

performance or goals. However, this assumption is, in practice, hardly proven true. People have various points of view and may see the same phenomena from two different, and at times conflicting, angles. When organizations are fragile and show little capability to adapt to changes (**agility**[24]), self-interest tends to take over to the detriment of the group, and thus eats away at group cohesiveness. People may try to trust each other and cooperate nevertheless, but that does not guarantee they will work on the same premises. It takes more than natural or forced trust and cooperation to ensure personalities, actions, and decisions work together. It requires the influence of KCFs. To that effect, **soft consensus** is one whereby group members reach consensus by feeling they are somewhat obliged to; **hard consensus** implies that the group members are fully committed to their decisions.

Of course, reaching consensus leads to value creation. In this context, the model of value creation deploys as in Figure 3.5.

This stylized diamond[25] reads as follows: Organizations set behavioral norms, which include good governance, a code of ethics, and internal as well as external rules of communication. Everyone has their own culture, and no one can escape the three managerial, measurable constraints that are the schedule[26] (by which all tasks have a beginning and an end), costs (which are capped to ensure the organization's financial future), and minimal standards of quality (designed to at least meet customers' and investors' expectations).

---

[24] Modern theory on project management expands, quite substantially, on agility.
[25] We do not mean to show statistical linkages between components; it is merely an illustration.
[26] As mentioned, sometimes named (•arguable error•) "delay," which is exactly what product managers want to avoid!

Lastly, organizations prepare contingency plans as they are subject to risks (external threat) and vulnerabilities (internal threats). Various organizations will handle these four managerial areas at various degrees and at various times, but they all resort to a combination of these *sine qua non* conditions managers must meet to ensure the long-term existence of projects and their deliverables. Out of these interplays emerge value, that is, the fact that organizations manufacture, produce, and/or offer something people want, cherish, or need, in short, the deliverable. Something that, in the end, improves the end users' living conditions.

Rule of Thumb: Risks and POVs must be jointly assessed.

Organizations that emphasize the three constraints of the iron (Bermuda) triangle work according to the classic model of project management. Those that focus on norms are likely to be in the education or military field, those based on value systems include not-for-profits, and finally, those that build their business on the unexpected include security services and emergency health care. This typology is often useful when preparing a business plan.

CHAPTER 3, CLASS EXERCISE #5:

Find a public case where the public, offering vastly divergent opinions, debated a project. Show and explain the absence of consensus using the proposed model on the KCFs.

PROPOSED QUESTIONS FOR DEBATE:
1. Is consensus always necessary?
2. Are there people, whether in organizations or in groups of allegiance, who are not suited for achieving consensus? Why?

A balanced mix of the four components of the diamond gives managers all the necessary tools to create value. Most organizations are, in fact, hybrid; yet, in the end, they tend to specialize in one of these four types. The ones of interest to us are, of course, those that are project management based. As discussed, we divide value into three forms, which standard marketing theory recognizes: perceived, added, and residual.[27]

---

[27] As discussed, after the 1950s, Toyota conquered the American market based on residual value. While its cars were more expensive than American cars, their resale value was much higher because they had a much longer "shelf life." Thus, in the end, the actual cost of the car was cheaper and customer satisfaction was higher.

As pointed out, the KCFs are not purely consequent to trust or cooperation between team members; they are part of the project's culture. Staff working in the spirit of unison foster the building of trust and further cooperation. Likewise, as trust and cooperation develop, staff are more apt at reaching consensus. However, trust and cooperation do not guarantee consensus. Some team members may well trust their teammates and they may work well together even while disagreeing, in a gentle manner, on certain aspects of a project. In general, as mentioned, trust is facilitated by the fact that team members share affinities, are benevolent toward each other, display high levels of competency, and are honest. Collaboration happens when people accept to be flexible and adaptive, exchange information openly, make an effort to solve problems together, and adopt a position where the good of the group surpasses the selfish interests of one.

We best illustrate the value KCFs bring by the tactical knowledge they encourage. According to the model, there are four types of acquired knowledge for the purpose of project management:

1. Know how to manage,
2. Know how to listen,
3. Know to express oneself, and
4. Know to prepare for the unexpected (see Figure 3.6).

CHAPTER 3, CLASS EXERCISE #6:

Based on your own experiences, discuss the models in Figures 3.5 and 3.6. Do you agree or disagree with their components?

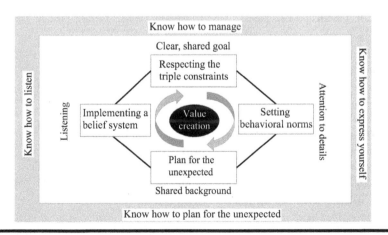

Figure 3.6 Preparing for the unexpected.

Note: A key aspect of project feasibility analysis is to attempt to guess what could go wrong.

PROPOSED QUESTIONS FOR DEBATE:
1. Are all people equal in terms of their capacity to manage, listen, express themselves, or anticipate changes to projects?
2. Are there cultures that are better than others are at gaining the knowledge discussed in Figure 3.6?

Since past projects have seen the production of a BOK, we must examine this knowledge under its four determinants (manage, listen, express, and prepare) in order to help project managers in the realization of similar yet unique projects being developed.

Note again that experts use KCFs in feasibility studies alongside an examination of KSFs and KFFs pertaining to the project being considered, whereas prefeasibility studies only deal with KSFs and KFFs (because they don't spend much time reviewing the human component of projects).

3.8 What Is a Causal Chain?

Projects are composed of inputs, a transformation process, and outputs. The inputs consist of the 4Ps (plan, processes, people, and power), which are idle at first but then interact with each other once managers engage in the transformation phase.

The transformation phase can be quite complex, and many software packages have invaded the markets to assist managers (e.g., MS Project). Yet, sometimes managers or students do not understand the concept of the **causal chain** well, although it is critical when assessing a project from a feasibility point of view.

Rule of Thumb: The causal chain becomes more precarious as intricacies between its constituents increase.

We explain causality by way of a kettle. If one puts enough heat underneath a kettle filled with water, given a certain atmospheric pressure, the water will boil 100% of the time. An instance where the water will not boil will never occur. Hence, there is causality. One hundred percent of the time, given the right conditions, the water will boil. There is just no way around it.

We now give the example of a cafeteria (Figure 3.7).

In this case, the best way to understand a causal chain is to work backward through each step, and ask the following question: "Would the present step exist 100% of the time without the presence of the previous step?" If the answer is "no," there is causality.

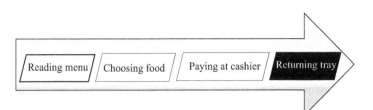

Figure 3.7   A cafeteria.

Note: In this figure, we show the flow of actions regular customers would adopt in normal circumstances (in all likelihood). While feasibility analysts must prepare for the unexpected, they must also remain reasonable so that the causal chain, such as the one exemplified above, makes (common) sense.

Let us be realistic and reasonable with our answers. (Of course, if one twists facts and logic, one can intellectually cheat the concept, but this will have little benefit in terms of developing the right mind for performing marketing feasibility analysis of projects.[28])

At the step where consumers are at the point of sale, would it be possible to pay for the products they have to choose if they had not chosen said products? Of course, the answer is "no" and the consumers will not get to eat. Therefore, there is a causal link: In order to pay for the chosen goods, the consumers have to choose the products first. You cannot pay for something if you have not chosen it. It is nonsense.

Let us now put ourselves at the step where consumers choose the food they desire. Could they choose the coveted products if the products are not there? Of course not! Therefore, having the food on display to allow clients to choose is mandatory in order for them to carry their trays to the cashier, where payment will take place. There is no other way around it, and this is another causal link.

Let us continue with this method. Can the staff place said products (foods) on the display shelves if they have not prepared them in advance? Of course not! Therefore, there is another causal link. Could the staff prepare the sandwiches if they have not received the bread, eggs, butter, etc. (the raw materials) that compose them (structural variables)? No! Could they cook hamburgers if there are no ovens? No. Could they refrigerate the soups if there are no functional refrigerators? No.

By proceeding in this way, analysts identify the causal chain and find exactly what they need in order to determine if project managers can materialize the project and the deliverables. It is as simple as that, yet this way of thinking escapes many books and teachings on project feasibility and on projects as a whole. In order to build a business, entrepreneurs and managers (often unconsciously) identify causal chains, whether in thought or on paper. Those who grasp their full sense and

---

[28] At times, this is something we see when teaching. However, the productivity of such behavior is nil.

potential are those who succeed. In fact, one can see causal chains as simply a series of points-of-no-return put forth: If you are at a point-of-no-return, it necessarily means that the previous steps leading to it were causal; otherwise, there would be a way of reversing the course of actions.

In preparing the causal chain,[29] marketing and project managers determine exactly what they need to succeed. Whatever is not within the causal chain but that is yet necessary will likely cause the system to fail. The concept is straightforward when dealing with mechanics, such as in the case of a corrugated paper machine: If the employee does not push the start button, the machine will not start! When it comes to human dynamics, however, things get quite complex. For example, we have just seen the KCFs, so let us choose one: Would people reach consensus if there were no clear, common, and mutual objectives? It is very tempting to say "no"—unless you imagine a unique scenario that occurs every ten million years and only under certain exceptional conditions. Obviously, even if people were to reach consensus without a clear, common objective, the project would not last very long. Very soon, the fragility of such consensus would surface and it would not be long before the group would have to meet again to work honestly, openly, and seriously. Hence, realistically (unless one wants to play devil's advocate beyond its useful role), there can be no consensus without a clear and common goal, by mere definition.

CHAPTER 3, CLASS EXERCISE #7:

Find an example close to your own experience, be it completing your studies or planning to become an entrepreneur. Lay out a causal chain by asking yourselves the right questions.

PROPOSED QUESTIONS FOR DEBATE:

1. Are there things, events, or people one cannot control when realizing a project?
2. To the best of your knowledge, would you say that causal chain thinking helps in preparing a feasibility study?

3.9 Conclusion

We have briefly reviewed what projects truly are, and we have outlined some of their key components (many of which are regretfully missing or disregarded in standard books on project management). Projects are not projects if they do not

---

[29] See the Ishikawa or fishbone techniques.

have a realization date; they are operations. Projects are not projects in the sense they require a feasibility study if they are duplicates of previous projects; obviously, if managers realized them before, they are feasible. Structural and functional variables, the iron triangle, the 4Ps, points-of-no-return, points of autonomy, KCFs, and causal chains are all concepts that assist marketing and project managers to gain a thorough understanding of the projects they work on. Posing the right questions and being realistic with the answers are quintessential behaviors in order to bring projects to a satisfactory completion.

We will discuss POVs in more detail in Chapter 4, which addresses project feasibility and, of course, marketing feasibility of projects.

3.10 Mind Teasers

Readers may use the mind teasers as questions in preparation for an examination or quiz.

1. Briefly explain
   a. the four KCFs and
   b. the four types of acquired knowledge used in project management.
2. Define and give examples of
   a. benchmarks,
   b. stage gates, and
   c. milestones.
3. Define
   a. points of autonomy,
   b. POVs,
   c. points-of-no-return, and
   d. soft consensus.
4. Differentiate between forces of production and means of production.
5. Draw a project's lifecycle, complete with necessary details.
6. Explain
   a. what a causal chain is by way of a simple example,
   b. what functional variables are,
   c. what structural variables are, and
   d. why "scope" is not appropriate as part of the iron triangle, from a perspective of project feasibility.
7. Give the four categories
   a. found under "people" (general staff, those who do not make strategic decisions for a project) and
   b. in which formalized knowledge, with its related documents, is organized.

8. List
   a. at least five of the ten management areas discussed in the PMBOK 6 and
   b. the eight characteristics and three functions that define a project.
9. True or False?
   a. In terms of structure and the means of production, "power" refers to the line of authority.
   b. KCFs are equivalent to KSFs and KFFs.
   c. Uniqueness and innovation are the same concepts.
   d. We categorize outputs into three sections: tangibles, intangibles, and nonedibles.
   e. Projects refer to undertakings that delve into past operations.

Chapter 4

What Is Marketing Feasibility of Projects?

The rebuilding of Notre-Dame can clearly be defined as a project. The deadline for completion is prior to the 2024 Olympic Games (to be held in Paris), comprising costs and norms of quality. The state is looking to create something unique that will attract even more tourists than before. The city—and the country—will have to market the new concept to a worldwide audience. We have to ask, Is it feasible?

4.1 Introduction

Project feasibility is about making sure that what promoters present to analysts are indeed projects and evaluating whether or not they are achievable. Determining the marketing feasibility of these projects entails glancing at their viability from a marketing (read, the clients') point of view but feasibility and viability are not one and the same. In this chapter, our discussion targets the field where project managers and marketing managers meet.

To recall, we define projects by the fact that all *sine qua non* structural variables are present, as well as at least two functional variables—nothing less! As mentioned, if the project is a close kin of a past one, there is no need for a feasibility study as it can be quite expensive; the success of the previous project has proven it is feasible.

Ensuring the feasibility of a project is not equivalent to defining whether it is viable, marketable, or profitable. To think otherwise is an "arguable error." A project is feasible when one can bring together the forces and the means of production as well as other resources as necessary, given the infrastructures in place, and then organize these for the benefit of the envisioned product to be brought to completion. Technically, this has nothing to do with a project's viability. We characterize projects by having a beginning and an end; to pretend to determine whether it is viable is close to an oxymoron. As a finalized deliverable, yes, it would be important to determine whether it is viable, but otherwise what's the point of monopolizing the forces and the means of production, resources, and infrastructures? Therefore, asking if a project is viable poses the question: Can we take for granted that the forces and the means of production will get us to the delivery date? That's certainly a question one can ask, but it's also a bit like asking, For how long will I be hungry until such time as I eat? In conclusion, let us avoid the term "viability" altogether.

As for a project's profitability, again, a project by itself is hardly profitable. It is surely deliverable and can and should be profitable (or at least break even). However, the project itself does not have to be profitable. Rather, we should regard any project as the cost of doing business. It becomes profitable, in a larger sense, only once it has actually delivered a product, provided a service, completed a process, or facilitated a research project, drawn conclusions, and collected money.

We offer here a very narrow sense of project feasibility to ensure readers distinguish the difference between the terms, which people sometimes use indiscriminately. There are, of course, eight different ways of looking at project feasibility (as previously discussed), and these entail, at least in part, some of these concepts. For example, doing a financial feasibility study is precisely about making sure we have the financing necessary to bring a project to completion, but also that the deliverable will be able to survive. Doing a marketing feasibility study is about anticipating end users' reactions; we bring everything we assume awaits us in the future and treat it as if it were the present. We then determine whether the project responds to a true, promising need-turned-opportunity; otherwise, there is no opportunity.

To some degree, the same logic applies to the other kinds of project feasibility studies, including environmental, legal, organizational, social, and technical.

As mentioned above, feasibility studies can be quite expensive, sometimes costing millions. To avoid such expense, we sometimes conduct **prefeasibility studies**. These take a global look at the project, but without going into too much detail. (In fact, many of the exercises we provide during our seminars are prefeasibility studies. They are a great way to train the mind to develop proper skills to complete prefeasibility studies. See Appendix 1)

Let us explore the fascinating subjects of prefeasibility and feasibility studies! First, we will spend little time discussing marketing, making the bridge between marketing and project management when necessary. In the section on strategic project management, we will take a closer look at marketing and show how marketers can use project-management tools to serve their own goals.

The learning objectives of this chapter are to understand what the marketing prefeasibility and feasibility of project are, to recognize what drives investigation into the feasibility of a project, and to connect marketing and project management.

4.2 What Are Prefeasibility Studies?

We adopt the following definition of prefeasibility studies:

> The prefeasibility study, which follows the initial value proposition of the project, offers a general view of said project using various analytical frameworks that allow the feasibility expert to make a recommendation on the suitability to conduct a feasibility study.[1]

To conduct a prefeasibility study, we use five analytical frameworks. These we describe further along in this chapter, but note we are not yet interested in knowing if the project is feasible—we are merely answering the question as to whether we should spend the time, effort, and money to do a full-fledged feasibility study.

We are not interested in scope (which is not measurable), or with customer satisfaction. In prefeasibility and feasibility studies, we must rather concentrate on what is measurable, or we will never be able to judge whether the project is feasible or measure its level of success. As discussed before, scope cannot be measured, and customer satisfaction is one of the most biased measures marketers use—one that serves the companies manufacturing the product but that actually says very little about the true feelings of customers. Let us put it this way: Imagine you are having a nice dinner on a Friday evening, with a person whose company you appreciate, and an anonymous person calls to ask whether you are satisfied with this and that. You say yes because you just want to get rid of the caller, politely if possible. In stark contrast, a calendar of tasks and activities is measurable: It has a beginning date, an end date,

---

[1] Mesly (2017).

and a set number of days. (Measurable!) Costs are, of course, also measurable. We set norms of quality against technical standards, each with precise measurements.

It is in that spirit we conduct prefeasibility and feasibility studies; shying away from this approach leads us down the wrong path. The same comment applies to making value judgments, seen in countless reports. Feasibility analysts sometimes try to take shortcuts, consciously or not, by relying on their own judgments and estimations. This is not a scientific approach and is why we must stick to facts and measurable factors.

To summarize, prefeasibility and feasibility studies require analysts to

1. Measure, measure, and measure and
2. Measure objectively.

It is often quite easy to unearth value judgments. We find them when analysts do not use supporting facts and available, reliable data, but rather invent or suppose they are there. They use expressions such as "it is interesting," or "it is important." Sometimes, would-be analysts transform themselves into professors of sorts. This is a no-no. The goal is not to teach; the goal of a prefeasibility study is simply, and merely, to make a choice between two options:

1. To go ahead with the feasibility study (→), or
2. Not to go ahead (↓).

It is as simple as that. Yet many of us are trained to emit judgments from a very young age without assuming any of the consequences whatsoever. We take for granted that our opinion is king of the universe. However, it is imperative analysts remain neutral; much like accountants do not discuss investments made by their clients, but simply work the numbers and sign their reports based on the data provided by these clients.

We cannot reiterate enough the importance of adopting the right posture. This is why we strongly encourage readers to complete the exercises we prepared, which should help to cure the tendency to make quick value judgments, or to make fallacious assumptions.

Prefeasibility studies are obviously upstream of feasibility studies, illustrated in Figure 4.1.

A prefeasibility study uses five frameworks of analysis. Try to look at them as if they were spotlights throwing light on an obscure object that may hide some hidden truths (or throw shadows around it), agendas, or defects. The five frames are as follows:

1. Definition (structural and functional elements),
2. Risks (commonly called risk analysis/risk assessment in project management),[2]

---

[2] This is a domain of project feasibility that, like financial feasibility, is well developed. There exist a number of excellent books on the subject.

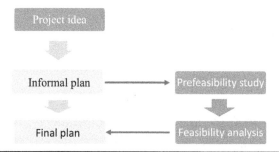

Figure 4.1 Flow of feasibility study.

Note: The above figure is a simple version of the flow of a feasibility study. Of course, many versions exist, but we aim to keep things simple.

3. Potential,
4. Parameters, and
5. Key success factors (KSFs) and key failure factors (KFFs),[3] which we discussed in Chapter 3.

One way to view these five frameworks is to see how they fit, figuratively, into the context of the eight areas of risk (or eight types of feasibility analyses), the 4Ps, and the iron triangle, as shown in Figure 4.2.

Figure 4.2 A stylized view at project feasibility.

Note: Many readers like visual elements, which can be a great mnemonic device.

---

[3] Note that in feasibility studies, which go over the five frames of analysis prefeasibility studies, KSFs and KFFs are complemented with the key consensus factors (KCFs) we discussed previously.

4.3 What Are the Five Frames of Prefeasibility Analysis?

Let us discuss in more detail the five frames of analysis used to achieve our goal of determining if a feasibility study is warranted.

4.3.1 Definition (Structural and Functional Elements)

The purpose of this frame is to ensure we have defined the project the best way possible. As mentioned, the main cause of project failure is inadequate definition, and it is thus quintessential to understand what the proposed project is all about. To assist with this, our seminars provide a fillable questionnaire, as given in Table 4.1.

Table 4.1 Definition

Project's name	
Goal	
Initial value proposition	
General description and norms of quality	
Calendar of tasks and activities	Beginning[a]: End:
Estimated costs (in *Euros*)[b]	
Main challenges	
Attributes	**Yes/No**
Structural Elements	
Responds to an opportunity?	
Includes stakeholders that are ready to participate?	
Is unique?	
Is innovative?	
Is subject to quality constraints	
Has a time limit?	
Requires financing?	

(Continued)

Table 4.1 (*Continued*)   Definition

Presents challenges?	
Is everyone of the above characteristics present?	
Functional Elements	
Generates deliverables?	
Produces knowledge?	
Has positive and/or negative impacts?	
Are at least two out of these three characteristics present?	

ᵃ We recommend to always identify the month by its name, not by a number, as different cultures place numbered months first, while others place days first. This may lead to confusion (e.g., 04/06 or 06/04; how to interpret the month?).

ᵇ When preparing their analyses, analysts should also always identify the currency and the date as the exchange rate changes with time, and those consulting the report may refer to different currencies.

The rule is that all *sine qua non* structural characteristics, and at least two out of three functional characteristics, must be present. If this is not the case, the proposed project does not deserve a feasibility study.

CHAPTER 4, CLASS EXERCISE #1:

Pick a project in which you have been involved and fill out the various tables provided in the chapter. Participate, if you wish, in our online exercises.

PROPOSED QUESTIONS FOR DEBATE:

1. Could having wrongly assessed the structural and functional variables of some infrastructure projects found in the marketplace explain their failure?
2. Do the structural and functional variables vary according to cultures/ countries/continents?

Table 4.2 continues with the analysis, in a longitudinal[4] way.

---

[4] In scholarly research, longitudinal refers to event 1 followed by event 2, with a single controlled variable inserted in between. There is a time effect that is not causal but simply due to the flow of time. The timeline can be very short (e.g., seconds) to very long (e.g., decades).

Table 4.2 A Temporal View of a Project

Phase	Main Tasks and Activities	Calendar (in days)	Costs (in Euros)	Norms of Quality
Vision				
Plan				
Mobilization				
Deployment				
Completion and evaluation				

Recall that a project is not complete until we evaluate it against three measurable criteria: calendar, costs, and norms of quality. Promoters who present projects without having carefully planned for this final evaluation are guilty of having a lack of understanding of their own projects!

4.3.2 Risks

This is commonly called risk analysis/risk assessment in project management.[5] It is standard in any feasibility study, just as a needs analysis is standard in any marketing feasibility study. Recall that risks come from outside a project, and whatever comes from inside is a vulnerability. Consequently, managers cannot control risks but can only anticipate and prepare for them. To think otherwise is an "arguable error."

Again, risks fall into eight categories, as per Table 4.3.

Table 4.3 A Quick Assessment of Risks

Type of Contextual Risk	4Ps	Comments
Financial organizational, (outside organization, not from the inside)	Plan	
Environmental	Processes	
Technological		
Marketing sociocultural, (outside, not from the inside)	People	
Legal	Power	
Political		

---

[5] This is a domain of project feasibility, like that of financial feasibility, that is well developed and for which there exists a number of excellent books.

We do not need to go into detail in a prefeasibility study; however, in a feasibility study, this aspect of the analysis requires much attention. Project-management theory identifies three types of risks that each has a corollary in the accounting statements for the project. These risks include the following:

1. What we can anticipate and prepare for (things known in advance),
2. What we must imagine and then weigh as to the likelihood of their appearance (these are guessed-at risks), and
3. What we do not know and what may catch us by surprise, hence the importance of the fourth element of KCFs (see Chapter 3).

4.3.3 Potentiality

We developed here an analytical tool to match the criteria that appear in Project Management Book of Knowledge [PMBOK]). Some elements may appear redundant, but it is better being safe than sorry—especially when it comes to expensive projects! (See Table 4.4.)

Note we don't assess people here because it is unlikely we would know who will be hired at this stage; we are only conducting a prefeasibility study. When doing a feasibility study, if the information is missing, we then recommend posing the same questions we asked for power, but this time for people.

CHAPTER 4, CLASS EXERCISE #2:

Pick a project you were involved in, even a personal one, and discuss its potentiality in hindsight, knowing what you know now of how it materialized.

PROPOSED QUESTIONS FOR DEBATE:
1. Is there a chance factor in the success of projects?
2. Are projects, from a certain point of view, about change?

Table 4.4 Project Potentiality

Questions on 3Ps	3 Ps	Yes/No
Is the business plan realistic?	Plan	
Is it well defined?		
Have change processes been carefully laid out?	Processes	
Are there clear measures of success?		
Does the management team have enough experience?	Power	
Are there measures in place to avoid derailment?		

Table 4.5 Project Parameters

Parameters	Comments
Define the KPIs	
Define the norms of quality of the deliverable	
Describe the need analysis (even a preliminary one)	
List the standards that must be respected (e.g., construction codes)	
Give examples of similar projects	

4.3.4 Parameters

Project prefeasibility and feasibility analyses are all about being objective, sticking to facts provided to us (if not, it is our duty to ask for additional information), and to measure that which we can. We outline this in Table 4.5.

It is very simple: If promoters and marketers have presented you, the feasibility analyst, with a document that is short on any (or all) of these critical measures, you know they have no way of measuring whether or not their project will be a success. Yet they'll often pride themselves on presenting the "best project ever," full of promises for a better future for everyone. It is important promoters and marketers show they have made an effort to compare their projects to ones that are similar. There is always something to learn, and perhaps by doing so many marketers would have a better definition of their projects, thus saving time, money, and efforts.

CHAPTER 4, CLASS EXERCISE #3:

You intend to finish the basement of your brand-new home in the city, a basement that is merely a cube of concrete right now. You want to build two bedrooms, a living room, and a washroom. List and describe the (real) parameters necessarily associated with such a project.

PROPOSED QUESTIONS FOR DEBATE:
  1. Can you afford to disregard city regulations?
  2. What are the consequences of not setting the right parameters upfront?

4.3.5 Key Success Factors and Key Failure Factors (KSFs and KFFs)

KSFs and KFFs were briefly discussed in Chapter 3, and academics often talk about them. Yes, KSFs and KFFs are important considerations, but are not absolute.

We must take a close look at them just to ensure promoters and marketers have considered all reasonable options; if not, they may not be serious enough to warrant doing a feasibility study (Table 4.6).

Many promoters and their marketers try to present their projects to potential investors in the most attractive light. Fair enough, but the analyst is the most

Table 4.6   KSFs and KFFs

Yes/No	Plan	
KSFs		
Dominant Strategy		
	Well-defined project charter	Clarity
	Access to financial resources	Efficacy
	Sound budget	
	Clear norms	
	Articulated calendar of tasks and activities	
Yes/No	Processes	
	Well-defined work methodology	Efficiency
	Reliable infrastructures	
Yes/No	Power	
	Experienced managers	*Competencies*
Triple Constraints	*Details*	*Will the Managers Respect the Constraints?*
Calendar	Beginning: End:	
Costs	Initially forecasted: Probable final costs:	
Norms of quality	Describe	
Contingency Strategy		*Yes/No*
Preliminary contingency plan devised		

(Continued)

Table 4.6 (*Continued*) KSFs and KFFs

Criteria	Present: Yes/No	Comments
KFFs		
Dreadful combination: poor plan and inadequate workforce		
Diverging interests among the stakeholders		
Blindness or over-optimism among the promoters and managers		

objective and pays factual attention to the project—not falling for smoke and mirrors.

Readers may see how exhaustive the angles of scrutiny are; the experts do marketers a favor by being so rigorous. Should the latter embark on a project with holes in it, they will suffer dearly in the end. They will have decimated their resources (sometimes personal) and achieved nothing but disappointment. The analysts must not be scared of offending a project's promoters and their marketers, as their duty is to amass and request all necessary information. They build their reputations on diplomatic elegance, certainly, but also on the rigor of their thorough analyses.

CHAPTER 4, CLASS EXERCISE #4:

Explain why dreadful combinations lead to failure, giving examples and discussing the various consequences of such a dynamic.

PROPOSED QUESTIONS FOR DEBATE:
1. When starting a project, can project and marketing managers ensure they don't sow the seeds of dreadful combinations?
2. Are there certain attitudes or work behaviors that nourish dreadful combinations?

4.4 What Are Feasibility Studies?

We have discussed prefeasibility studies and, in the process, touched on feasibility studies as well. We adopt the following definition of the latter[6]:

---

[6] Inspired by Mesly (2017).

A project feasibility report is a comprehensive study that examines in detail the five frames of analysis of a given project in consideration of the 4Ps, its risks and POVs, and its constraints (costs, calendar, and norms of quality) in order to determine whether it should go ahead (→), be redesigned (←), or else be totally abandoned (↓).

In addition to using and upgrading the five frames of analysis of the prefeasibility study the feasibility study uses six more tools, known as the 6Ps of strategic project management.

A feasibility study goes in depth, using five frameworks and the strategic 6Ps, in order to determine if the project is truly feasible. In particular (and as briefly mentioned), this is when we use the KCFs, which we add to the frame of KSFs and KFFs because people play a vital role in feasibility studies as opposed to prefeasibility studies. We take a magnifier and focus more on the core of a project. First, we decide what kind of feasibility analyses are needed (e.g., marketing and financial). Then, we look at these five frameworks in a general light, as in the prefeasibility report. We then pay attention to the strategies outlined by the 6Ps.

One can perform a quick assessment of the feasibility of a project by simply asking three primary questions:

1. Does the project manager have managerial skills?[7]
2. Is there mutual trust among stakeholders?
3. Are the necessary resources available?

Although these three questions are simple, one will soon discover they are actually hard to answer. Only a thorough feasibility analysis allows us to answer them with the necessary rigor. However, if you instinctively say no to at least one, then it is a good indication something is about to go wrong.

In a feasibility study, as opposed to a prefeasibility study, the notion of value becomes incremental. We like to look at a project and its deliverables in the long term. Recall how we defined **value** (Figure 4.3).

Here, again, we consider value from the point of view of project management's iron triangle. In what way does the project bring value, whatever its form, by improving the timeline of the project *versus* past similar projects? In what way does the project provide better cost control *versus* past similar projects? Does the project adhere to stricter norms of quality *versus* past similar projects?

Answering this from a marketing feasibility analysis point of view indicates whether the project delivers, and what it promises to deliver. If it does not, it will not respond to needs or opportunities. Thus, it is not feasible in the sense that it fails to keep its word.

---

[7] Project management literature focuses heavily on technical skills but much less (alas!) on the relationship-building skills of project managers (Creasy and Anantatmula, 2013).

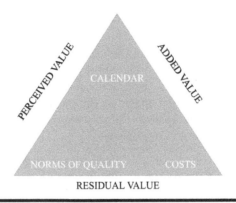

RESIDUAL VALUE

Figure 4.3 Values, revisited.

Note: Prefeasibility studies simply want to answer the question as to whether a feasibility analysis is necessary; the latter is usually quite expensive. In marketing feasibility analyses, the analysts want to ensure the deliverable will bring value to the end users.

We place much more emphasis on needs assessment or, as we shall see, opportunity assessment, as previously discussed. A project has two audiences: the investors, who will only be pleased if the feasibility analysis is convincing, and the end users (some of whom participate in a co-creation effort toward the project), who are satisfied only if they see and experience the value the deliverable offers. Hence, a marketing feasibility study is a feasibility study that focuses on satisfying the projects' investors and end users. It is one that ensures end users will find value in the promise the project, by its nature, provides.

We have a number of analytical tools at our disposal to conduct feasibility studies, so let us review them in the sections that follow.

4.5 The Six Tools Used in Feasibility Studies

Prefeasibility studies use five frames of analysis. We complement the last one—the KSFs and KFFs—by KCFs, which can only be assessed during feasibility studies. This is because we examine people with more scrutiny in a feasibility study. (In a prefeasibility study, rarely have people and power been determined.) For example, the feasibility analysts will meet with the project managers, and maybe (if possible) some relevant staff. Hence, they will gain an idea as to whether the emerging team is likely to bond and work well together. Feasibility analyses add another set of analytical tools (six to five frames of prefeasibility analyses). These, which we consider strategic project management tools, are as follows:

1. The 4Ps (in more detail): plan, processes, people, and power;
2. PRO: pessimistic, realistic, and optimistic scenarios;

3. POVs: points of vulnerability (under extensive scrutiny);
4. POE: point of equilibrium;
5. POW: product, organization, and work breakdown structures; and
6. PWP: work psychodynamics.

Together, we call them the 6Ps of strategic project management and feasibility studies. We have already discussed some, such as the 4Ps, but here we delve into more detail and focus on how they can assist in the strategic management of projects. These tools shed a different light on projects, with the objective of unearthing whatever could jeopardize them. They assist managers who are concerned with minimizing or managing

1. Risks (often related to calendar),
2. POVs (often related to costs), and
3. Errors (closely related to norms of quality).

We represent this entire strategic dynamic in Figure 4.4.

There are also a number of other analytical tools in project management, many of which have filled numerous books over the decades, and many of which have been developed by the army to please engineering minds.[8] They may be incorporated in the 6Ps; for example, you can find some in the process component of the 4Ps (plan, processes, people, and power) in the list given below. We invite readers

Figure 4.4 Strategic project management at a glance.

Note: Strategic project management equates with the six tools used in the feasibility analysis. Recall that we also proposed six elements for strategic marketing as well: innovation, segmentation, positioning, differentiation, targeting, and building loyalty.

---

[8] There are many other tools available, such as FAST (functional analysis system technique), FMEA (failure mode and effect analysis), FTA (fault tree analysis), and OPA (optimal path analysis), which inspired our viewpoint.

Table 4.7 Other Analytical Arsenal and POVs

Phase	Possible Analytical Tool	Helps Finding POVs'...
Vision	Analysis by comparative tables	Identity
	Analysis using PRO scenarios	
	Analysis using decision tree	Presence
	Analysis of probability of risks	Importance
	Multi-criteria analysis	
Planning	Cause-to-effect analysis	Identity
	GANTT, PERT	Presence
	Critical path analysis	Importance
	Optimal path analysis	
	Sensitivity analysis	
Mobilization	Cause-to-effect analysis	Identity
	Critical path analysis	Importance
Deployment	Optimal path analysis	Importance
	Critical path analysis	
Completion and evaluation	Critical path analysis	Importance
	Cause-to-effect analysis	Identity

to consult other books and scholarly articles to learn more. Experienced feasibility analysts are usually very well versed in them, especially those who work for companies that provide this kind of service (see Table 4.7).

Let us now review the 6Ps of strategic project management. We do not pretend the following surpasses existing tools; we simply attempt to enrich the field of project management with tools that most likely appeal to marketing managers, who are not necessarily trained in engineering.

4.5.1 The 4Ps of Project Management

We have already discussed the 4Ps of project management and showed how they connect with the 4Ps of marketing. Therefore, we will spend little time here discussing them, but wish to complement our understanding with tools to which marketers can easily relate. The objective, again, is to bridge the gap between marketing and project management. In feasibility studies, people become a focus of analysis, which is not the case in prefeasibility studies because we are instead looking at the project before we start considering who, exactly, will likely be the stakeholders.

> Simple definition: A **project's culture** is determined by the 4Ps: plan, processes, people, and power.

4.5.1.1 Plan

Plan is about anything that gives a taste of the future; it is of things managers have not yet realized. Instead, we prepare to put deliverables into the market; that is, we prepare for the marketing of these deliverables. Quite often, project managers are trained engineers. They use charts, flows, and complex diagrams (PMBOK provides many charts that can become quite confusing, in fact). Marketers need to find a way to talk the same language as their clients, which they do by making their messages as attractive and friendly as possible. Here are some possible tools.

4.5.1.2 Summative Triangle

The summative triangle is one way to talk to the clients who participate in a project, or to end users and media, so people have an understanding of how the project is evolving (see Figure 4.5).

The summative triangle has a powerful illustrative function. Each circle represents a phase of the project, starting from the vision phase at the center and expanding, over time, toward the outside. We separate the circles into three equal

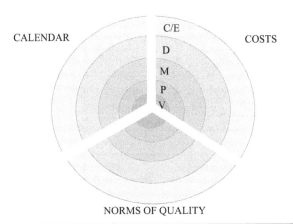

Figure 4.5   The summative triangle.

Note: This tool is useful when working with clients and the media, because it is simple and easy to adjust over time.

sections, one each for timeline (calendar), costs, and norms of quality. Managers can fill in the figure with actual numbers as things develop. For example, they may put in a value of USD 50,000 in costs during the mobilization phase. Marketing and project managers can post the summative triangle (which is also a circle!) on a wall, on the Internet, or on any media that permits simple, effective communication to the targeted audience. For the feasibility analyst, it is also a tool for grasping, at once and in the simplest form, what the project looks like overall.

CHAPTER 4, CLASS EXERCISE #5:

Take either a real or invented example and prepare a summative triangle.

4.5.1.3 Critical Points

We discussed critical points in Chapter 3; they all have a meaning when it comes to marketing. For example, marketers can use stage gates to launch a communication strategy and get the stakeholders excited; indeed, the completion of a project deserves a special social event, drafted to make every stakeholder proud of the accomplishment as well as of present and future benefits. To this day, people still celebrate the first man to go into space or walk on the moon. We like to summarize the critical points as in Table 4.8 because marketers can actually list, in a separate column, what activities they plan for each point.

Note POVs are not critical points in the purest sense; marketers cannot take advantage of them to prepare a communication strategy. In fact, it's quite the opposite!

4.5.1.4 Structural and Functional Elements

As mentioned throughout this text, defining a project is turning it into a success. In Chapters 1 and 3, we explained how to evaluate whether an element (or variable) is structural or functional. We define any object fully by both its structural and functional variables, and nothing less. Missing one of the two is only reaching half

Table 4.8 The Marketing of Critical Points

Name	Type	Symbol	Detail
Benchmark point	Benchmark measure	■	Point of no-return
Stage gate	Stage	♦	Five phases of the lifecycle
Milestone	Key achievements	●	Start point
			Final point of delivery

the definition.[9] Descriptive variables are timeless and not polarized; they simply are what they are.

Structural and functional variables belong to bonds called descriptive variables and do not take timeline (calendar) into consideration. There are three other types of bonds between variables (or elements of a project), and managers should be aware of all four types:

1. Descriptive,
2. Influential,
3. Longitudinal, and
4. Causal (briefly discussed in Chapter 3, in the section covering causal chains).

Let us review each one.

4.5.1.4.1 Descriptive

Descriptive variables are the ones marketers like the most. They transform them into expressions end users can relate to. If the **structural variable** (S) of a project deliverable is a sturdy wheel, the marketers may find using this terminology is not appealing to the masses. If they reposition this variable with marketing savvy, however, they could say, "A wheel that makes your head turn!" (Or something like that!)

If the **functional variable** (F) of this wheel is that it spins at 130 km/hour, the marketers, in an attempt to raise the end users' interest, could conceive of something like this:

"The wheel that's ahead of time."

Readers can see how useful, proper and descriptive definitions are; both marketing and project managers benefit from this initial effort. A marketing feasibility of projects analyst would be thrilled to see a project proposal that seizes this opportunity to prove, at least in part, that it is feasible.

Hence, marketers and project managers work hand in hand to define a project in the best and most complete way possible. From this, marketers extract attributes end users seek and label them in a way that will entice them to buy or use the deliverable once ready.

We saw examples of descriptive variables in Chapter 1, and we review some of them in Table 4.9.

4.5.1.5 Non-Descriptive Variables

Influence variables (I) are a bit more complex and have more to do with strategic marketing plans and project management. There are two types: direct and indirect. **Direct bonds** simply say that one variable (the independent) influences another

---

[9] For more, see Mesly (2015a).

Table 4.9　Examples of Descriptive Variables

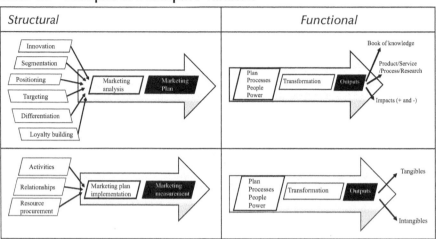

(the dependent), either positively or negatively. The models we saw in Chapter 1 illustrate there are many direct bonds of influence. Figure 4.6 shows some.

In Figure 1.14, a number of variables exert a direct influence. A negative perception does not necessarily lead, over time, to a lack of trust; yet a negative perception of a deliverable would most likely reduce any trust end users would otherwise have in that deliverable. As trust gains ground, end users are more likely to cooperate with the managers of the deliverable as there is a positive influence (I^+). Let us look at another example: When clients participating in a co-construction effort within a project feel they can trust the marketing and project managers, they are more likely to cooperate. This is a positive bond of influence (I^+). As people cooperate better, they are more likely to commit to working together for both of their benefits from the

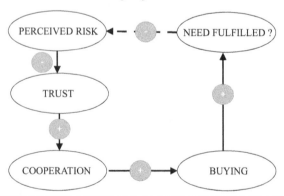

Figure 4.6　An example of variables exerting a direct influence.

Note: Analysts must take the effort to understand how things relate, including by way of direct influence.

project. When they both benefit, there is no reason to develop a negative perception of the project or its team members. If the latter happens, the influence is negative (I^-); when commitment or benefits go up, negative perception goes down. As for the sense of fairness in the model (represented by the scale), it is an indirect variable.

Indirect variables come in two forms: **mediator** (I^+ or I^-) or **moderator** ($I^±$). Fairness is a mediator as it acts as a bridge between two variables (trust and cooperation; the independent and the dependent variable). Moderators, as the name indicates, moderate the relationship between two variables, and this influence may be either positive or negative ($I^±$); it depends on the circumstances. We saw a moderating variable ($I^±$) in Chapter 1, as shown in Figure 4.7.

Marketing and project management, hopefully, have a positive influence on satisfaction by fulfilling the consumers' needs; but there are always people who are not happy whatever we do. Therefore, we cannot guarantee the influence will always be positive, and we cannot see that their actions push needs toward satisfaction. Rather, we are happy if at least 50% of the end users are satisfied to the intended level, as the product fulfills their needs.

Why are these technical elements important to marketers? Because it is a great advantage to know what influences consumers' behavior. It is crucial to differentiate what is a straightforward influence *versus* what takes more input to sway end users in the desired direction: appreciating and using, or buying, the deliverable. Furthermore, many conflicts arise because people don't see the difference between a moderating variable and others. Recall that a moderating variable can have a dual effect: It can be positive for X% (e.g., 55%) of the total population and negative for 100%–X% (e.g., 45%) of the same total population. Often, people argue they are right and their *vis-à-vis* are wrong. Well, they are both right. If they are debating over a moderating variable, such a variable can have either effect; therefore, no need to quarrel! Take aggressive publicity, for example. For, say, half the population, it is a source of motivation to get off the couch and rush to the store and buy whatever is on sale and advertised as such. For the other half, however, the same will really annoy them and they may swear they will never buy from the retail outlet employing such aggressive advertising. In this particular case, marketers must weigh which side is more likely to generate profits; in all cases, one side will be less happy than the other.

Figure 4.7   An example of moderating variables exerting a moderating influence.

Note: Misunderstandings over moderating variables often lead to conflict, as stakeholders do not realize they can both be right even if they seem to express opposite views.

This kind of trade-off occurs all the time, and there are many such situations with projects. In fact, you can think of the three elements of the iron triangle as moderators of sorts: If you please the accountant by reducing costs (by buying cheaper materials), you displease the quality control guru. If you cut short the delivery date, you please the investor because return on the money comes faster but, again, you may do so at the detriment of quality.

As mentioned, **longitudinal bonds** refer to a clear time effect. If you ask someone to stand on one foot, for example, with time that person will eventually want to put the raised foot down. We don't know when that will be, but it is fair to assume they will not turn into a flamingo.

The last, but not least, type of bond is the **causal bond**, which is also affected by time but more so by the causality linking two variables; they can also be positive (C^+) or negative (C^-). We have seen marketing and project managers must understand the concept of causality (causal chain). As well, they must also understand that, for one cause, there can only be one effect, and for one effect, there can only be one cause. This is the best way of thinking in order to create a linear flow of events, much like a sentence starts at one point and ends at another. If your way of thinking includes more than one cause for a single effect, or more than one effect for one cause, then the model may be good but not articulated in a way that feasibility analysis likes to operate. With the concept of one cause one effect, we are always certain of what is happening (100% of the time). This is a true causality chain and allows for better control of the entire project and of its marketing plan.

PMBOK identifies a certain number of causal bonds, which we include in Table 4.10. We added how POVs relate to the type of causal bond.

Table 4.10 The Bonds Between Elements of a Project or its Marketing Plan

Bond name	Code	Impact on POVs
Longitudinal	(T)	Moderate impact on POVs when not in a critical stage
Longitudinal loop back	(t)	
Direct influence	(I^+) or (I^-)	May affect process elements, but not dramatically
Indirect influence		
Mediator	(I^+) or (I^-)	May affect process elements, but not dramatically
Moderator	$(I^\pm)$	May cause managerial problems
Causal	(C^+) or (C^-)	May have a devastating effect on projects

(Continued)

Table 4.10 (*Continued*)   The Bonds Between Elements of a Project or its Marketing Plan

Type of Causal Bond	Critical Level of POVs
Not ending the previous activity (C^+) → not starting the next activity	Low because no costs are associated with the new activity yet
Not ending the previous activity (C^+) → not starting the current activity	Moderate because both activities are assumed to be near their end
Starting the previous activity (C^+) → staring the current activity	Moderately high because the processes will not work unless both activities are given the go-ahead
Starting the previous activity (C^+) → ending the current activity	Critical because there is intense pressure once the entire process has begun

It takes a fair bit of time to familiarize oneself with the various bonds (descriptive, influence, longitudinal, and causal) but, eventually, one realizes how useful this way of thinking is and how it simplifies life, especially when dealing with complex projects. Feasibility analysts can use these concepts to better define projects, or direct project marketers to do just that.

4.5.1.6 Processes

It is time to define more precisely what inputs go into the transformation phase, illustrated In Figure 4.8.

Many feasibility studies forget to examine the elements that compose the "P" of processes, so let us discuss them briefly. In prefeasibility studies, we take only a quick look at the "P" of people but, in feasibility studies, we pay much more attention to them as we know POVs are mostly related to people. The "P" of processes is another example where we dedicate a fair bit more time. We divide processes into two subsections: resources and means of production (forces of production go into people). In resources, we have energy sources such as furnaces and generators; infrastructures such as roads, hospitals, electrical/gas networks, and so forth (which inexperienced analysts tend to forget)[10]; money (the financing aspect of a project); and materials. In means of production, three elements are commonly found in accounting documents: building, equipment, and machinery. A fourth element, which is also often forgotten (an "arguable error"), is measuring instruments. As explained, feasibility studies are all about measurement. These include the identification of key performance indicators (KPIs), forms (e.g., specification sheets, or a Material Safety Data

---

[10] Infrastructure is an important consideration in less-developed countries. In advanced countries, analysts take them for granted; yet they must be cognizant of them.

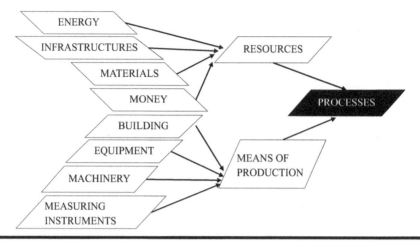

Figure 4.8 Processes, better defined.

Note: Regretfully, many models found in project management books fall short on the inputs.

Sheet [MSDS]), and actual measuring instruments. (If they are not there, how can one possibly verify the standards of quality are respected?)

4.5.2 PRO

Upon doing the exercises during our seminars, participants will notice there is a section where we assume they have listed all that could go wrong with their project. Doing feasibility studies, both technically and from a marketing point of view, demands that the analysts (or the marketers) do this exercise for every project.

PRO refers to pessimistic-realistic-optimistic scenarios. Most promoters stick to "optimistic," and many tend to neglect the other two; in fact, one of the major causes of project failures is an unrealistic calendar of tasks and activities. Never mind being optimistic, they are bluntly unreal, unachievable, and impossible—like asking a giraffe to dance the tango with a hippopotamus. However, from a feasibility analysis point of view, the most important of the three kinds of scenarios is "pessimistic." Feasibility analysts have to play devil's advocate; they have to identify anything that can go wrong, and they have to question this repeatedly. What if end users shy away from the deliverable? What if the deliverable has a major defect that turns end users off? What if there is a change in the government's social support of the project?

The technique to devise a pessimistic scenario is as follows:

1. Form a small group of people, ideally all stakeholders in the project in one way or another, with different points of view (this works best with no more than nine people [or an odd number] in case a majority vote is needed);

2. Brainstorm whatever comes to mind with respect to what can go wrong (within reason[11]), but do not attend to sort, categorize, or order ideas just yet;
3. From this list, try to combine similar scenarios or regroup different scenarios into one that encompasses them reasonably well;
4. Narrow this down to the ten events that would most likely go wrong; and
5. Sort them hierarchically.

Once complete, review whether the marketers or project managers have put plans in place to address these potential problems. If not, assume the project is not feasible. Remember that anything that can go wrong *will* go wrong; that is usually the nature of the beast. It's better to be safe than sorry, especially when human lives are at stake—as is the case with many large, expensive projects.

CHAPTER 4, CLASS EXERCISE #6:

Pick an old or existing project and brainstorm on everything that can go wrong. Let yourself loose!

PROPOSED QUESTIONS FOR DEBATE:

1. Are there projects for which everything goes as planned?
2. How useful it is to imagine scenarios that are truly out of proportion (e.g., preparing for an extra-terrestrial invasion of your project's site)?

A standard calculation for PRO (also discussed in PMBOK) is to assume the pessimistic scenario weighs one-sixth, the realistic one-sixth, and the optimistic four-sixths of the total scenarios. However, allotting one sixth for a doomed scenario, we find, is rather generous; it means there is roughly a 17% chance of the project failing. In other words, you, as a marketer, would present the project to some investors and tell them there is a nearly 20% chance they will lose their investments (but, you would add, "Trust me, the project is great, excellent, and promising!"). What are the chances any investor would be thrilled by such an opportunity to dig a financial hole?

We much prefer to use the following formula, which we find more pragmatic and more in line with our own estimation of the percentage of POVs present in any project (about 4%–7%) (see Equation 4.1).

■ Equation 4.1: Proposed project's potentiality

$$\text{Project potentiality} = 0.04\,\text{Pessimistic} + 0.70\,\text{Realistic} + 0.26\,\text{Optimistic}$$

---

[11] Yes, there are exceptions. Charles Pathé, a great innovator in the film industry, suffered a catastrophic fire at his plant late in the 19th century. This was after having bet heavily on the market, against all odds, and caused him near bankruptcy. Chances of a like chain of events are slim.

Put differently, when marketers and project managers conceive projects, they must estimate three routes. Often, the mere act of thinking what could go wrong helps project promoters actually improve their projects, or else helps them come up with innovative solutions that add overall value to their initial offers. Being optimistic also presents benefits. By clearing one's mind of the nitty-gritty details, one can rise above the facts and imagine something that would otherwise be overlooked—like looking above the trees instead of being stuck in the swamp. Promoters and marketers can henceforth conceive an ideal scenario they will then propose to end users to entice them to use or buy the deliverable.

As such, the PRO tool enhances the project without being overly expensive. When doing the exercises during our seminars, participants should pay particular attention in making this PRO effort. In fact, it is often quite fun to start with, as people tend to let loose and generate what could turn out to be valuable ideas and well-articulated contingency measures.

4.5.3 POV

As mentioned, in the purest sense, POVs are not critical points like benchmarks, milestones, or stage gates; we'd rather get rid of them. Another way of looking at prefeasibility studies is to focus on POVs (briefly discussed in Chapter 3). Recall **POVs** express the conditions of vulnerability by which any of the 4Ps (plan, processes, people, and power) of a project, or a combination thereof, makes the project susceptible to failure. For example, recall the infamous Alcatraz, the jail just off the coast of San Francisco, and the much-publicized escape of three prisoners on June 11, 1962. These three inmates had experience in escape; so, this was a prefeasibility study (or pilot test) of sorts. Alcatraz was a maximum-security penitentiary from which no one had ever escaped. However, the inmates became aware of one POV in the heavily guarded structure: a disabled fan vent on the roof, which they could access through an unguarded, meter-wide utility corridor that ran alongside their cells. They managed to access it after chiseling through the small vent separating this corridor from their cell walls. The point here is that the two vents did not represent a risk, but they certainly represented a POV from a security-system perspective. The true risk was outside of the system, *per se*: the frigid water the three prisoners attempted to cross with a makeshift inflatable raft; the cold water is not a POV, and hence it is a risk. This is simply an example, but we hope it illustrates the difference between POVs and risks.

Keeping POVs in mind is crucial in any prefeasibility or feasibility study. Both studies may or may not make use of prototypes but, generally, prototyping is a necessary step in complex and/or novel product development for both engineers (project managers) and marketers. Analysts take a hard look at the project and adopt the attitude of a devil's advocate, trying to find anything that could go wrong (well, anything within reason that would command us to consider the possibility). Of course, there could be a tsunami of Higgs particles submerging the project at once and, of course, this remains in the realm of possibility, but we would rather remain reasonable and pragmatic!

Recall that **POVs** are "temporal and physical points along the various phases of a project that impede the calendar, costs, and/or quality of the project as it faces adverse conditions (negative forces) [and] whether these conditions are under human control or not."[12] In fact, POVs have the most significant impact on costs. This, too, should be in the back of our minds at all times.

In Table 4.11, we provide an overview of low- and high-vulnerability contexts.[13]

We composed Table 4.11 by extracting information from over 100 projects, at different times (read, decades), in different sectors of activity, and in different cultures. Low-vulnerability contexts indicate projects are more likely to be feasible than high-vulnerability contexts. In all cases, managers must take corrective measures as early as possible. In fact, project marketers and feasibility analysts and funders can use Table 4.11 as a checklist before contemplating any form of funding. From a marketing perspective, the same criteria apply: A marketing plan spoiled with internal conflicts and lack of commitment from top management is likely to fail. To achieve the goal of building customer loyalty, one fosters careful, realistic actions.

Table 4.12 crosses POVs and the three constraints of the iron triangle.

As the readers can judge from Table 4.12, managers are not empty-handed when it comes to evaluating POVs. Complaints received from customers, or even from future end users (e.g., pressure groups that do not want to see the project succeed), open access to a gold mine of information. Astute managers will now

Table 4.11   Low- and High-Vulnerability Contexts

Low-Vulnerability (Internal) Context	High-Vulnerability (Internal) Context	4Ps
Appropriate resource commitment	Poor resource commitment	Plan
Carefully planned changes, solid results	Rapid changes, quick results	
Realistic expectations	Unrealistic expectations	
Set responsibilities	Ambiguous roles	
Operational changes	Strategic changes	Processes
Shared vision	Conflicting perceptions	People
Supportive top management	Uncommitted top management	Power

---

[12] Mesly (2017).
[13] Mesly (2017).

Table 4.12 POVs and Their Constraints—Examples of Measurements

For Output Items	For Time	For Costs	For Norms of Quality
Forms processes	Cycle time	Budget variances	Complaints
Items assembled	Equipment downtime	Contingency costs	Defects
Productivity	Late reporting	Cost by account	Error rates
Sales	Overtime	Delay costs	Number of accidents
Units produced	Response time to complaints	Overhead costs	Rejects
Work backlog	Repaid time	Penalties/fines	Rework
—	Work stoppage	Unit costs	Scrap
—	—	Variable costs	Waste

better know what problems are likely to occur, or what are currently weakening the project. The key here is to measure. As we have stated a number of times, feasibility analysts stick to what is measurable and develop tools to do so, if required. There is no place for random judgments or biases, which often hide underneath the level of conscientiousness. Because complaints come from current clients, or potential future end users, they form a bridge that connects marketing and project managers. This is one of many examples whereby the two managers can work together and can learn from one another.

In Table 4.13, we position POVs along a project's lifecycle. Technically, there are no POVs during the vision phase, but this does not mean one cannot run the above-mentioned Table 4.12 as a quick check. The planning phase is, in large part, about defining the project. Hence, POVs relate mostly to descriptive variables, which encompass structural (S) and functional (F) variables. In the middle section of Table 4.13, we highlight the fact that costs associated with fixing POVs increase substantially as the project nears its end. This is because the more we move forward with a project, the more units of plan, processes, people, and power have been invested and become an integral part. Recall the causal chain and the example of the cafeteria. If we were to change the list of items purchased from an existing one to one that does not comprise products currently in stock, this means we have to go back and change everything that takes place before getting to the cashier. We have to revise the presentation platform where products are waiting to be picked up, the menu list, the preparation done in the back kitchen, the reception of new raw

Table 4.13 POVs Along the Five Phases of the Project

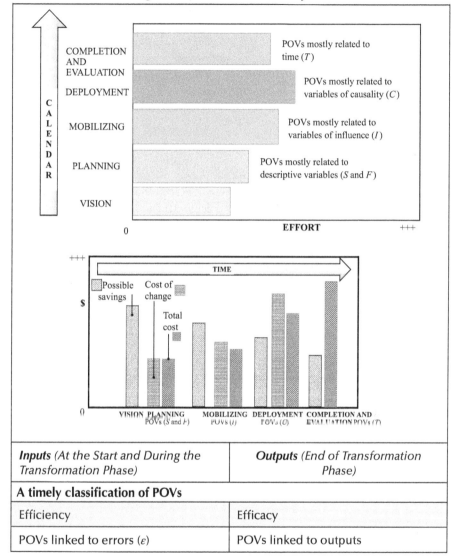

Inputs (At the Start and During the Transformation Phase)	*Outputs* (End of Transformation Phase)
A timely classification of POVs	
Efficiency	Efficacy
POVs linked to errors (ε)	POVs linked to outputs

materials, and so forth. This is why it is so important to identify and tackle POVs early on, and why feasibility analysts must be experts in doing just that.

The bottom part of Table 4.13 takes another look at POVs from a longitudinal point of view, but emphasizes the fact that, early on, POVs are intimately related to errors, including calculation or measurement errors, ordering mistakes, and the like. **Efficiency**, or the ability to use all inputs to produce the envisioned output (the important deliverables), is of the essence here. As time goes on, POVs increasingly relate to **efficacy**; this refers to how close the deliverable compares to its initial design. During a project, POVs negatively affect efficiency early on and, later on, efficacy. In the end, marketing and project managers aspire to a project that presents both efficiency and efficacy, with each one of these two elements being measurable. In other words, measuring efficiency and efficacy is one sure way of determining if there are POVs in the transformation system—which a project is.

Feasibility analysts must sit down with promoters and review the potential POVs, using the various concepts explained above.

This sums up our discussion on POVs; let us now discuss their conceptual twins.

4.5.3.1 Risks

As we discussed in Chapters 1 and 2, risks are vastly different from vulnerabilities. They are external to the project, and we categorize them into eight sectors. Project management literature and courses discuss at length the notion of risks, sometimes without recognizing these eight factors and often without delimiting the frontier between risks and vulnerabilities. This is an "arguable error."

As a way to stylize their relationship, we like to formulate the following equation when comparing risks and vulnerabilities (see Equation 4.2):

▪ Equation 4.2: Vulnerability

$$\text{Vulnerability} = \frac{k}{\text{Risks}}$$

The value (k) is a constant we determine based on various factors.

We can look at risks from yet another angle: a managerial point of view. As the standard theory goes, experts must

1. Document and deal with them;
2. Check whether they are active, and thus represent a hazard; and
3. Continuously monitor them.

We'd say the same kind of managerial attention applies to POVs.

4.5.3.2 SVOR

SVOR stands for strengths, vulnerability, opportunity, and risks. It is the equivalent of SWOT (strengths, weaknesses, opportunity, and threats),[14] which we find incomplete and which hardly applies to project management, or marketing for that matter. There are no such things as threats, but rather risks.

---

[14] In military contexts, psychology, marketing, or ecology, one talks of perceived threat, but not in project management *per se*. In standard project management theory, one talks of risks, not of threats. We do a risk-assessment analysis, not a threat-assessment analysis. Originally, SWOT was developed by the military.

Rule of Thumb: A project is not feasible if the positive forces (strengths and opportunity) are fewer than the negative forces (risks and vulnerabilities).

The SVOR system is a nice way of identifying which elements are internal (strengths and vulnerabilities) and which elements are external (opportunities and risks). It also identifies which are positive (strengths and opportunities) and which are negative and could jeopardize the project (risks and vulnerabilities) (see Table 4.14).

A good rule of thumb is as follows: A project is likely to succeed if the strengths are in line with the opportunities, given that the vulnerabilities and their corresponding risks are low. Thus, the SVOR system is a handy way of guessing whether or not a project is likely to succeed. It is, naturally, included in any feasibility study, and appears as one of the core measurements in the exercises provided during our seminars (See Appendix 1).

We go into more detail with respect to SVOR in Table 4.15.

Note the role of infrastructures in the SVOR table, an element managers sometimes ignore when resorting to SWOT analysis. This highlights further why using

Table 4.14   The SVOR System

	Internal	*External*
Positive forces	Strengths	Opportunity
Negatives forces	POVs	Risks

Table 4.15   A Detailed SVOR Table

	Internal	*Hypothesized Link*	*External*
Positive forces	Strengths	Strengths given constraints = Infrastructures/Opportunity	Opportunity
Negatives forces	POVs	Risks given constraints = k/POVs	Risks
Or:			
Positive forces	Strengths	Strengths given constraints = Infrastructures/Opportunity	Opportunity
Hypothesized link	POVs = 1/Strengths	k	Opportunity = 1/Risks
Negatives forces	POVs	Risks given constraints = k/POVs	Risks

SWOT as a feasibility analysis tool is an "arguable error." Infrastructures are crucial elements of construction projects. One needs to evaluate whether proper roads, electricity, hospitals, telephone lines, and so forth are in place and are functional. This may seem of little significance when working in advanced countries or near major urban centers, but it certainly plays an inescapable role in less-advanced territories.

In the SVOR system, what makes a feasibility analysis a marketing concern is the research done on opportunity. As mentioned previously, the opportunity is to the seller what the need is to the buyer (or end user of the project's deliverable). A sound marketing feasibility analysis certainly includes a needs assessment, but it should also include an opportunity assessment as need and opportunity go hand in hand when it comes to projects. Projects are the expression of the need; they are the opportunity in some narrow sense. Hence, a marketing feasibility analysis is essentially a feasibility analysis that includes a well-developed need/opportunity analysis.

CHAPTER 4, CLASS EXERCISE #7:

Draw an SVOR table for a project you know well.

PROPOSED QUESTIONS FOR DEBATE:

1. Are there projects where no infrastructures would be considered?
2. Can POVs be so omnipresent that they jeopardize even superior strengths and opportunities?

4.5.4 POE

POE stands for point of equilibrium. A well-balanced budget, adjunct to an articulated and realistic timeline and proper norms of quality, plays in favor of projects. It is easy to relate to this concept and get a feel for how to interpret it by resorting back to the PRO scenarios and SVOR (Figure 4.9).

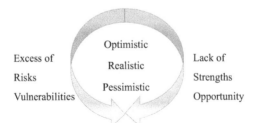

Figure 4.9 Point of equilibrium.

Note: POE is a rather abstract concept, but it can be useful when identifying excesses and shortages (which are always related to project difficulties).

According to this model (or conceptual tool), project equilibrium sits between excess of risks and vulnerabilities and lack of strengths and opportunities. We cannot realistically measure POEs; we can, however, attempt to minimize risks and vulnerabilities and maximize strengths and opportunities. Nothing is perfect in real life, and there is a constant interplay between excesses and shortages. This applies to both the marketing and project management fields. To put it differently, projects that fail have lost foot; they never found their POE. There were too many excesses and shortages; managers spread their efforts too thinly, resulting in an unbalanced project.

CHAPTER 4, CLASS EXERCISE #8:

Pick a project you know well and describe what elements displayed excesses and shortfalls.

PROPOSED QUESTIONS FOR DEBATE:

1. Do projects with no excesses or shortfalls exist?
2. Should end users be blamed for shortfalls that affect projects?

4.5.5 POW

Another tool feasibility analysts may use is that of POW, an acronym for product, organization, and work breakdown structures. Product breakdown structures (PBSs) or product hierarchies (family of needs → family of products → class of products → product range → product type) are well known in marketing under the name product tree (project managers not familiar with the concept can find a description in almost any marketing book). An example of a product tree is a flying object, which is then classified as one with wings, which is thereafter sub-classified as one with engines. Therefore, we know it is neither a bird nor a parachute: It is a plane. We can decompose the plane according to its structural and functional elements, and each part or sub-part will have specification sheets. Spec sheets are the best way to ensure customers cannot claim to be disappointed. If the deliverable does what we promised according to the spec sheets, there are no reasons to complain. Work Breakdown Structures (WBSs) are simply the lists of tasks and activities. Organization breakdown structures (OBSs) refer more specifically to organizations, while WBS is a standard term used in project management.

In Table 4.16, we propose how to measure each one of these breakdown structures.

The POW is simply another way of looking at a proposed project. As readers can judge, it includes elements we have seen with other angles of analysis. This double checking of a project's feasibility is a way of eliminating potential mistakes

Table 4.16 A POW and Its Measurements

Item	POW
KPIs are identified	PBS
Deliverables are well defined	
Management of resources is well planned	OBS
Dominant and contingency strategies are in place	
Budget is complete and realistic	
Organigrams are crystal-clear	
Tasks are explicitly laid out	WPS
Calendar of tasks and activities is set	
Work has been forecasted	

in our own work. Project managers may want to complement this table with further measurements, depending on the complexity of their project. Marketing managers, or the clients who evolve with the project, find here a quick way to feel comfortable—providing the answers are satisfactory. We can look at it this way: A good understanding of the deliverable (a prime concern for marketers and their clients), handled by a sound organization that runs work processes efficiently, is likely to generate positive results.

CHAPTER 4, CLASS EXERCISE #9:

Pick a project you know well, even a personal one, and prepare its POW the best you can.

PROPOSED QUESTIONS FOR DEBATE:

1. Do all projects require an organization to manage them?
2. Is PBS more important than OBS or WBS?

4.5.6 PWP

PWP treats work psychodynamics. We have already seen a model to that effect, in Chapter 1 (see also Appendix 2). This is one field of study that clearly differentiates prefeasibility from feasibility studies. In the latter, some stakeholders, such as the project managers, have already been identified and may have started to work together. Some clients, who are likely going to be involved as co-creators of a project, may also have been identified. Hence, people plays a crucial role at this point, including how stakeholders interact from a psychological point of view. This is what we call work psychodynamics. This field has been vastly studied, and countless books and articles exist on group dynamics, personalities, and so on. A review of psychological constructs that appear often in project management literature suggests the following commonly found psychological constructs (Table 4.17).

This understanding has led to a model we mentioned in Chapter 1, and which we present once more (Figure 4.10) to discuss core competencies.

Table 4.17   Common Psychological Constructs in Project Management Literature

Theme	% of Authors
Trust, including acceptance of others, affinities, technical competencies, conformity, integrity, loyalty, open mindedness, reliability, security, sensitivity to other cultures, team spirit	29
Cooperation, including adaptation, collaboration, common goal, communication, flexibility, proactivity, reciprocity, sharing, team training	25
Control, including feedback, follow-up, leader commitment, leadership, performance, project definition, resource allocation, task planning	21
Commitment	15
Fairness, including impartiality, neutrality, win-win	9
Distance (cultural)	2

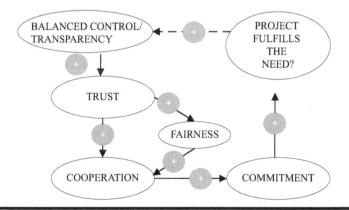

Figure 4.10   The links between the essential psychological constructs.

Note: Here, we specify with more precision the core psychological model.

As we have seen numerous times, in project management managers have to have a certain level of competency. We also discussed the importance of trust, and Figure 4.10 pinpoints what core competencies are mandatory in projects. Put differently, the feasibility analysts, when ready, should meet and interview the stakeholders who will work for the proposed project (people) or direct/manage it (power). The six core competencies they must have are outlined in Figure 4.11 and summarized as follows.

For people:

1. The capacity to trust, and to generate trust,
2. The capacity to cooperate, and
3. The ability to commit.

For power (notwithstanding the above three competencies):

1. The capacity to control,[15]
2. The capacity to be transparent, and
3. The capacity to be fair.

As discussed previously, trust is usually determined by considering four criteria: affinities, benevolence, abilities (competencies), and integrity. "Competencies" is the easiest component of trust to measure: CVs, letters of recommendation, reputation, and so forth attest to one's fit with the tasks required. Experts usually express cooperation by displaying flexibility (the capacity to adapt to change), exchanging

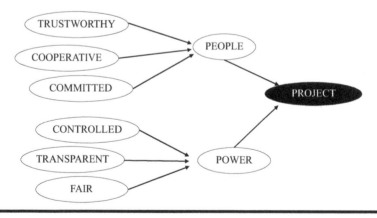

Figure 4.11   The six core competencies of project management.

Note: Our research shows these six competencies are essential in about any project, small or large, whatever the type of project or the country in which it takes place.

---

[15] Dulewicz and Herbert (1999) have shown that taking charge and being assertive, as well as having the willingness to win and the dislike of losing, are traits that further assist managers in their organization.

information, finding resolutions to common problems (closely linked to the KCFs seen previously), and bonding (displaying interest in others). The easiest way to evaluate the level of cooperation is by measuring conflicts *versus* outputs.[16] Many conflicts and poor results indicate that little cooperation has taken place. In fact, while trust facilitates the achievement of goals, conflicts affect one critical aspect of the iron triangle: They inevitably delay projects.[17]

The problem is that one must wait until some action has taken place between various stakeholders to measure cooperation. Therefore, among all the factors that best indicate whether there will be a good fit between the various stakeholders before commencing a project are measures of competencies. Expert can do this along five dimensions (Figure 4.12):

1. Experience, of course,
2. Skills,[18]
3. Talent,
4. Personality as an instrument of social skills,[19] and
5. Work ethics (an angle of analysis that is often forgotten).

Certainly, scoring high on all five parameters ensures having a strong, reliable workforce. This cannot be assessed in prefeasibility studies; it truly belongs to feasibility studies and should be included in about every kind of study, including marketing, or when referring to marketing stakeholders, such as marketers and participating clients.

Figure 4.12 Measuring competencies.

Note: As mentioned, we cannot do a feasibility analysis worth the name without measurements.

---

[16] We refer to Taylorism, a well-known work engineering process that separates workflows into small tasks that require little decision-making on the part of the workers.

[17] Idrissou et al. (2013).

[18] This ties in with the notion of project capability, seen previously (Davies & Brady, 2000).

[19] As Consoli (2006, p. 77) mentions: "the underlying cause of most disputes was the reaction to the '…personalities of the key players…'." Competencies must therefore be evaluated preferably in terms of fit of personalities among the key stakeholders.

CHAPTER 4, CLASS EXERCISE #10:

For a project of your choice, position yourself in terms of the six core competencies.

PROPOSED QUESTIONS FOR DEBATE:
1. Can competencies be acquired through education and experience?
2. Are some people more naturally gifted with respect to the six core competencies than others?

Team members will achieve little if they cannot trust each other, play (nasty) games, and have hidden agendas. They must cooperate, stick to the plan, join efforts, and commit to success by paying attention to details. Managers must exert control and provide enough transparency to instill confidence. (Nowadays, except in certain countries, dictatorial management enjoys little support.) However, too much transparency may jeopardize a project, as competitors may sniff out the opportunity to steal an idea or potentially patentable technology, or adverse groups may identify a weak point (the vulnerability) and attempt to disrupt the flow of tasks and activities (see the notion of **techno-predation** at the conclusion of this book). Finally, lack of fairness inevitably first makes people disappointed, then angry, and then feeds what experts call **counterproductive work behaviors** (CPWB). These behaviors interfere with the workflows and the team atmosphere. They include such behaviors as overt hostility,[20] hiding or distorting information, stealing documents or prototype samples from colleagues, falsifying documents, fostering conflicts, and the like. We consider anything that impedes on the six core competencies of project management as, in essence, counterproductive.

> **Rule of Thumb:** The more the wrangling among stakeholders that is intense, frequent, and cover critical issues, the less feasible the project is.

It is useful for feasibility analysts, and for marketing and project managers who work together, to identify among their staff (including among clients who are invited to participate in the realization of the project) who is currently, or likely, to become a disruptive force, and who is likely to foster conflicts.

We find it useful to consider four elements on the possibility of conflicts, which we refer to as structural components: (1) an opportunity, (2) a means to express it, (3) a motive, (4) and some form of obstacles. People rarely enjoy conflict for the sake

---

[20] As we have seen previously, hostile people "...simply exploit whatever resources are available to gain advantage..." (Graham, 1996, p. 68). These people are "pursuing one's concerns at the expense of the other party" (Wood and Bell, 2008, p. 130).

of conflict. Rather, their thrill comes from the challenges they face and overcome, making their adversaries upset. People may diverge in opinions; this does not equate with conflicts. Disagreements are normal and can be productive, but conflicts, as we express here, have a negative effect on projects. People who engage in conflict seek not the realization of the project, but their own satisfaction; they derive satisfaction from overcoming (or torpedoing) well-reasoned challenges, for example, when vehemently arguing to make sense of a particular situation (see Figure 4.13).

A typical scale for conflict[21] is given in Table 4.18.

Generally, within projects, especially with larger ones where a large range of personality types may be involved, one will find the kinds of personalities we discussed previously. Stable people, those who can balance various traits and coping mechanisms according to situations, are the most suitable for facing the daily hazards that quickly fill a project's routine. Overly aggressive, defiant people (who say "no" for the pleasure of saying "no"), or those who hesitate or may be solitary (antisocial), are all sources of potential problems within groups. Experienced feasibility analysts will easily identify these people by doing some group exercises[22]; they must necessarily examine people and power by interviewing stakeholders, doing group

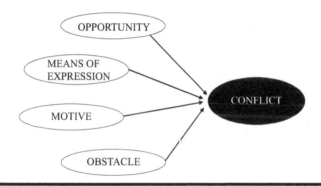

Figure 4.13   Structural components of conflicts from a MFP point of view.

Note: Better understanding the components of a conflict means being better equipped to deal with them before they jeopardize the project. Conflicts are POVs.

---

[21] Based on the works, among others, of Ohlendorf (2001), Creasy and Anantatmula (2013), and Van Goozen et al. (2000).

[22] There are a number of tools to that effect, but it is beyond the scope of this book to delve further into this subject. In particular, conflict resolution techniques, such as accommodating, avoiding, collaborating, or competing, have garnished the literature for decades (e.g., Kilmann and Thomas, 1975; Wood and Bell, 2008). Readers must note that feasibility studies, unlike prefeasibility studies, pay much attention to PWP because, ultimately, projects are created and operated by humans. As discussed, POVs are the making of humans, either directly (e.g., toxic interactions between team members) or indirectly (e.g., through a badly conceived machine).

Table 4.18   The Barometer Leading to Aggression

Aggression	Conflict zone
Provocation	
Frustration	
Indifference	Indifference
Random trust	Comfort zone
Conditional trust	
Blind trust	

exercises, and even distributing personality tests (e.g., the Myers-Briggs personality test—MBTI).[23] Nothing prevents them from seeking the advice of a work psychologist when doubtful about a project.

At our end, we, as authors and analysts, have always completed our analyses with interviews of the promoters, planned project managers, and, when possible, other staff. We often discover quite a large gap between the document promoting the project and stakeholder's reality: When challenged, some promoters change their story or twist facts. Often, an over-optimistic outlook on the project hides the fact the promoters have not done their work scrupulously, even though this is essential. Feasibility analysts can be quite annoying that way, first because they throw a dark cloud over a project that was, according to its promoters-turned-marketers, the best thing since pasteurization and antibiotics and, second, because the analysts unveil weaknesses right in the face of excited promoters. Few people like to see their balloon punctured with the emotionless needle of scrutiny. Yet this is the job to do here, and in the end, feasibility analysts do promoters-marketers a favor: They save them time, effort, and money by outlining what is wrong and what can go wrong.

Feasibility analysts can verify whether the following conflict countermeasures are in place, either formally or informally, for example, by interviewing the project managers (see Table 4.19).

Most conflicts occur over either priorities or procedures. We present a summary of the differences between good and conflict-ridden work in psychodynamics terms in Table 4.20. The feasibility analysts, when interviewing the project's promoters, may find inspiration from this table and drive the discussion along some of its components. For example, a question could be "Have you checked on the local infrastructures?" or "Suppose there is a disagreement between you, the promoter-marketer, and the project manager you intend to employ; how would you deal with it?"

---

[23] Research results tend to support the utility of personality assessment tests to a limited degree (Barrick, Mount, and Judge, 2001), yet they do not have strong predictive power.

Table 4.19   A Checklist of Conflict Countermeasures

Phase	Source of Conflict	Checklist	4Ps
Planning	Priorities	Plan well defined?	Plan
		Joint decision-making?	
		Contingency plan?	
		Focus on goal?	
	Procedures	Detailed procedures?	
		Schedule respected?	
Mobilization and Deployment	Priorities	Technical measurements from the start?	Plan
	Procedures	Tasks and priorities defined?	
		Feedback favored?	Power
		Collaboration encouraged?	
		Work in progress (WIP) supervised?	
Completion and evaluation	Priorities	Reallocation plan prepared?	Plan
	Procedures	Stakeholders informed?	People, process, power
		Problems solved promptly?	

At the time of the project proposal, promoters may not yet have chosen all employees. They may not have even yet contemplated whom they are likely to hire, or would want to hire. Project analysts prefer to know, as much in advance, whom the project is likely to involve. As noted by a scholar, "… based on the interviews with project management professionals, it is clear that project-based work does produce a number of unintended pathogens that can significantly affect both the way in which project management professionals do their work as well as the manner in which they interact with critical stakeholders."[24]

The above knowledge is useful for both marketing and project managers. While working together, they themselves will evaluate each other on their competencies

---

[24] Pinto (2014, p. 385).

Table 4.20 A Functional, and a Dysfunctional, Work Atmosphere

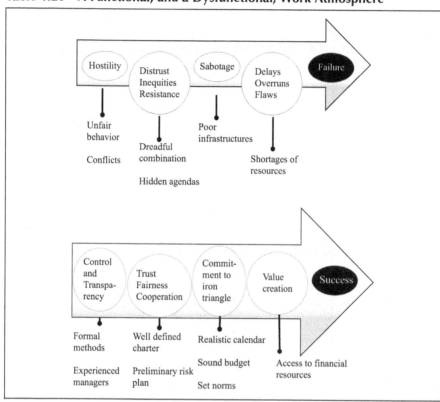

and capacity to deal positively with conflicts. Chapter 5 provides plenty of examples, from projects all over the world, that exemplify how important work psychodynamics are in projects, highlighting the fact that feasibility analysts must pay attention and try to anticipate what will happen as the project unfolds.

CHAPTER 4, CLASS EXERCISE #11:

Pick a project in which you have been involved (even group homework), and in which conflicts occurred. Discussed how this affected your level of motivation and desire to complete the project.

PROPOSED QUESTIONS FOR DEBATE:

1. Are there people more geared toward conflict than others?
2. Do conflicts arise because people put their ego ahead of the common goal?

4.6 What Are the Particularities of the Marketing Feasibility Analysis of Projects?

As discussed, marketing feasibility does not limit itself to a needs assessment, which consists of measuring the level of need among the target markets. We also want to make sure we would be justified in transforming these needs into opportunities.

With respect to the target market of end users (as opposed to investors, who would be interested in financial data and marketing information for the project itself), recall from the introduction that marketing projects is about asking six questions:

1. Does the project aim at putting an innovative product (deliverable) in the market?
2. Will the product (deliverable) answer the needs of the intended market segments?
3. Has the product (deliverable) been measured against its competitors for its lifecycle, and against its family of products?
4. Have managers put forth a sound strategy with respect to the product (deliverable) itself, its price, its promotion, and its intended distribution?
5. In what way will the consumers/end users perceive the value, and different attributes, of the product?
6. How will repeated consumption (or use) of the product (deliverable) be guaranteed, assuming such repetition will take place?

To answer these questions, we simply need to prepare a marketing plan that goes through the six components of strategic marketing, keeping in mind there are two objects under consideration: the project itself and the deliverable:

1. Innovation: for example, identifying the type of deliverable (product, service, etc.) and describing the needs along the project's dimensions (e.g., real or latent; see below), as well as the promise (e.g., attributes, benefits, sacrifices, value) and the product's unique features (e.g., design, functionality);
2. Segmentation: for example, identifying the type of market in which marketers will position the product (e.g., normal market, oligopoly, market share) and the end users' profile (e.g., consumers or industrial buyers; anxious buyers; occasional, regular, loyal, etc.);
3. Positioning: for example, delimiting the product's appeal against competitive products, using such tools as perceptual maps, and identifying all the market agents and their relevant roles (e.g., regulators and regulations);
4. Targeting: for example, identifying a small group of opinion leaders for the launch phase and discussing the 4Ps to the level of scrutiny required for the analysis;

5. Differentiation: ensuring the product stands out and has appeal; for example, by celebrating every milestone reached along the project's lifecycle and along its regular lifecycle;
6. Loyalty building: assuming the deliverable is not short-lived, defining the kind of relationships envisioned with end users (e.g., relational, customer relationship marketing management [CRM]-based), and planning programs to ensure end users will remain loyal to the product (the deliverable); it also includes discussing the consumption experience from beginning to end (e.g., perceived threat or risk, compulsive buying).

More particularly, and with respect to needs, the feasibility analyst must address the following:

1. The type of need or motivation, for example, a latent desire;
2. The sector of economic activity it covers, for example:
   a. Clothing;
   b. Communication (e.g., portable phones);
   c. Education;
   d. Food, alcohol, or tobacco;
   e. Health and beauty aids (HABA);
   f. Shelter (house and house maintenance);
   g. Sport and entertainment; and
   h. Transportation.
3. The factors influencing demand, namely,
   a. The availability of substitutes,
   b. The type of need (as located along Maslow's hierarchy of needs),
   c. The level of utility,
   d. The budget constraints (or the capacity to buy the coveted product),
   e. The urgency of the need (or how pressing it is, represented by present *versus* future consumption),
   f. How well the product quantitatively and qualitatively responds to the need, and
   g. The opportunity to buy and consume said product (the opportunity set).

Our seminars offer templates (including PowerPoints) to assist participants in preparing this important aspect of marketing feasibility analysis. Collecting data requires qualitative and/or quantitative research, and accessing primary (first-hand) or secondary information. Any basic marketing course or book will cover these in detail. (We do not discuss research methods in the present book; we refer readers to Mesly, 2015a.) As marketing feasibility analysts, we have to remain consistent while using the standard marketing tools within the realm of project management. As long as marketers can prove the value of their initiative, there are few reasons why a feasibility analyst would not recommend that the project go ahead. As seen previously, marketing feasibility of projects analysts must determine whether the project proposal is marketable and, ultimately, whether the deliverable the project will produce is marketable. A marketable project proposal entices potential investors to fund the project; it requires answers to the question as to whether the deliverable is marketable. Does the marketing plan unveil a need and an opportunity? Has it identified a profitable, attractive market? Are the real psychodynamics of the end users (including the clients who will participate in the creation of the deliverable; therefore, in the project itself) captured? More specifically, have the marketers of the project carefully considered

1. The attractiveness/utility of the proposed innovation?
2. The segment of consumers who will find more benefits than sacrifices in the project?
3. The position of their offering against competitive or other offers across time?
4. The target segments and appropriate product/price/promotion/place policies?
5. The development of a differentiated approach?
6. The planned-for long-term relationships with the community and the end users?

Answering "no" to these questions is saying "no" to the project or, at least, begs for improvement. Potential investors are unlikely to invest in a project if it is not feasible or, better said, proven feasible.

As for the layout of a marketing plan prepared specifically for the project and its deliverable, it could look something like this (Figure 4.14).

The step named "marketing presence strategy" is precisely about the above-listed six components of marketing management. Marketers, and possibly project managers, should discuss how they will implement and measure market presence. The better and more measurable the marketing feasibility analysis is, the easier it will be to find an investor. At all times, the marketing feasibility analysis, unlike occasional and general marketing plans, should stick to the three elements of the iron triangle. These are the best way to measure projects and their deliverables (and hence, customer satisfaction) as they take into account timeline, costs, and

Figure 4.14   A marketing plan layout.

Note: This is simply one of many types of marketing plans.

Table 4.21 Intuitive Marketing Movers, Blockers, and Delayers

Movers	Blockers	Delayers
Proper innovation	Lack of patents or legal protection	Unrealistic timeline
Proper segmentation	Poor selling infrastructures	Excessive costs
Proper positioning	Intense or illegal competition	Inability to apply, control, and enforce quality standards
Proper targeting	Diluted markets	—
Proper differentiation	Lack of competitive advantage in relation to the key attributes	—
Proper loyalty-building strategies	A volatile consumer base	—

norms of quality. Unlike other marketing plans, those designed within a marketing feasibility proposal should include an analysis of marketing movers, blockers, and delayers. To recall, see Table 4.21.

CHAPTER 4, CLASS EXERCISE #12:

Update the marketing plan you prepared earlier, taking into consideration all of the above elements.

PROPOSED QUESTIONS FOR DEBATE:

1. What could you do to make your plan as convincing as possible?
2. Why do certain stakeholders act as blockers or delayers?

4.7 Conclusion

While the title of this book concerns the marketing feasibility of projects, we first discussed prefeasibility studies. These rely on five frames of analysis to obtain a big picture of the project and to determine if a feasibility study is necessary or not. The latter can be quite expensive; otherwise, the prefeasibility study may suffice to start a project with a limited scope. Projects that are more complex usually require feasibility studies, which look at the five frames with more depth, and which use six tools, known as the 6Ps or the six strategic managerial tools.

Marketing feasibility of projects is a feasibility study in the purest sense, one whereby analysts put a particular emphasis on needs/opportunity analysis. We want to ensure the promoters can effectively market their projects to potential investors and end users. Potential investors are not going to invest in a project if they are not certain the project appeals to intended end users.

Increasingly, project managers involve selected clients in the development of deliverables (products, services, etc.). This means project managers are no longer engineers or technical managers by trade; they engage socially with various stakeholders and, in particular, clients who co-develop with them. Innovation, nowadays, is a multi-panel effort, and marketing and project managers must obligatorily open lines of communication.

The best way to prevent conflict is to adopt standards and measures. One cannot argue much with what is measurable: Project promoters, marketers, and the clients, as well as project managers, can only look at the numbers objectively and verify whether what they planned was actually achieved or not.

This chapter provided some insights to that effect. The marketing and project managers may decide how to go about the five frames of analysis and the six tools of strategic project management, and they may want to develop their own calibrating systems, such as those that measure experience, knowledge, talent, skill, work ethic, and personality while dealing with the "P" of PWP.

The exercises we provide during our seminars attempt to train participants to develop an objective mind; they do not cover all aspects of prefeasibility and feasibility analyses, as this would be overwhelming in terms of pedagogically sound training. Rather, each exercise allows readers to reinforce their capacity to read project proposals as neutrally as possible, while pointing out their potential sources of failure.

If anything, this book, and the exercises, should help improve the mutual understanding between marketing and project managers, an effort ultimately benefiting end users. A better work atmosphere, and a closer fit between marketing and project management objectives, can only serve the interests of end users, for whom marketers and project managers conceive and realize projects, one way or the other.

Chapter 5 discusses in more detail the unique relationships that take place between marketers, clients, and project managers. Using the marketing and project management knowledge presented in this book, and by doing a needs/analysis (for which countless books exist), is to the advantage of these stakeholders. Everyone will feel more confident the deliverable is of value; pride takes over selfish interests, common goals prime over hidden agendas, and success prevails over failure.

As a final note on this chapter, recall that everything is possible within reason if stakeholders define their projects as completely as possible and carefully prepare for the unexpected. Throwing an analytical light from various angles on the same object (the project) ensures nothing comes from seemingly out of nowhere and jeopardizes its realization, thus making everyone happy. There is no reason to be unhappy when people work together to create value for the entire community. Going through the steps we have devised in this chapter allows the analyst to determine whether the project should go ahead (→), be amended and improved (←), or else abandoned (↓), as regretful and heartbreaking as this may be. Feasibility analysts are after fact-based information and seek to render a decision that, ultimately, saves effort, time, and money.

Better safe than sorry.

4.8 Mind Teasers

Readers may use the mind teasers as questions in preparation for an examination or quiz.

1. Briefly describe
   a. the five phases in the lifecycle of a project and how they relate to the iron triangle (costs, calendar, norms of quality),
   b. the steps to devise a pessimistic scenario,
   c. the three types of bonds that exist between variables (other than causal), and
   d. the two types of risks (besides known risks).
2. Draw
   a. an example of a summative triangle and
   b. the three zones of the barometer leading to aggression in a work environment.
3. Explain
   a. the difference between SVOR and SWOT and
   b. what a dreadful combination is.
4. Give
   a. at least four questions to assess whether there are conflicts among team members,
   b. examples of at least two CPWB,
   c. the proposed project's potentiality mathematical formula,
   d. three examples of common technical tools used in the project management field (other than GANTT charts), and
   e. two examples each of low- and high-vulnerability contexts.
5. List, and briefly, describe
   a. four of the KSFs,
   b. the 6Ps of project management used in feasibility studies,
   c. the five frames of analysis of a prefeasibility study, and
   d. three KFFs.
6. List
   a. at least four common psychological constructs found in project management literature;
   b. the eight categories of risk, associating them with the relevant Ps of project management; and
   c. the six core competencies sought in project management.

7. True or False?
    a. A feasibility study proposes one of three options: to go ahead with the feasibility study ($\rightarrow$); to go ahead, conditional to changes being made and re-evaluated ($\leftarrow$); or not to go ahead ($\downarrow$).
    b. A good understanding of the deliverable (a prime concern for marketers and their clients), handled by a sound organization that runs work processes efficiently, is likely to generate positive results.
    c. A plan must be realistic and well defined.
    d. A project feasibility report is a comprehensive study which examines, in detail, the five frames of analysis of a given project, with consideration of the 4Ps, its risks and POVs, and its constraints (costs, calendar, and norms of quality) to determine whether it should go ahead ($\rightarrow$), be redesigned ($\leftarrow$), or else be abandoned ($\downarrow$). In addition to using and upgrading the five frames of analysis of the prefeasibility study, the feasibility study uses six more tools, known as the 6Ps of strategic project management.
    e. A project's culture is determined by the 4Ps: plan, processes, people, and power.
    f. A quick assessment of the feasibility of a project can be made by asking three primary questions: Does the project manager have managerial skills? Is there mutual trust among stakeholders? Are the necessary resources available?
    g. Efficiency is the ability to use all inputs to produce the envisioned output (the important deliverables).
    h. Ensuring the feasibility of a project is equivalent to defining whether it is viable, marketable, or profitable.
    i. KPIs must be identified ahead of launching the actual project.
    j. Managers are concerned with minimizing or managing risks (often related to calendar), POVs (often related to costs), and errors (closely related to norms of quality).
    k. Marketing feasibility of projects is a feasibility study in the purest sense, one whereby analysts put a particular emphasis on needs/opportunity analysis.
    l. POVs are critical points identical to benchmarks, milestones, and stage gates.
    m. POW is an acronym for product, organization, and work business structures.
    n. Prefeasibility and feasibility studies require analysts to make objective measurements.
    o. PRO refers to pessimistic-realistic-optimizing scenarios.
    p. Project equilibrium sits between excess of risks and vulnerabilities, and lack of strengths and opportunities.
    q. Structural and functional variables are polarized (+ or −).

    r. The feasibility study, which follows the initial value proposition of the project, offers a general view of the said project using various analytical frameworks that allow the feasibility experts to make a recommendation on the suitability to conduct a financial study.

    s. To assess fit, experts can use five dimensions: experience, skills, talent, personality, and work ethic.

8. What is

    a. a *sine qua non* condition and

    b. the difference between risks and vulnerability?

Chapter 5

What Do Marketing Experts, Project Managers, and Clients Have in Common?

A building undergoing reconstruction in Luxembourg. Only the skeleton remains. For successful realization, projects must sometimes be revisited, and conditions reset. (This is partly why feasibility studies exist.)

5.1 Introduction

Increasingly, marketing practice is moving away from viewing clients as "targets" to become instead relational human assets that can be cultivated or, better yet, leveraged.[1] This is because customers now, to a high degree, influence the response to project developments. Furthermore, as mentioned in the introduction, sometimes the team members assigned to a project know each other and have worked on projects in the past, often in the same company they work for. In other words, in these circumstances, the forces of production do not arrive on the playfield of the new project empty-handed: They have acquired some knowledge about the other team members. That is a bonus for the feasibility analysts because they hence have a more accurate understanding of one of the 4Ps (people).

By nature, some industries are more project based than others, such as is the case with information technology (IT) development. A growing number of companies work based on projects, and as they build their business models, they must establish solid relationships with their clients. Not only are clients sensitive to the specifications/attributes of the products they purchase (their design and functionality), but they are also sensitive to their timely delivery. This is especially true in a world where the likes of Amazon dictate the service modality to potential clients. Clients as participants in the project are thus one aspect of people that the feasibility analyst should consider when addressing the 4Ps of project management.

Research has found that communication and coordination between marketing teams and project-related technology teams differentiate projects. Otherwise, these teams, whether overly conservative or technically focused, may become embroiled in conflicts. On the other hand, where harmonious relationships prevail, a project's success rate improves by nearly 50% over dysfunctional teams. According to many project managers, customers' needs rank as their third priority, just after design and planning. Other considerations follow, but none is as crucial as these first three: design, planning, and customers' needs.[2] Hence, customers' needs, and their underlying psychological frameworks, appear to be fundamental in marketing and project management the world over. (For more, revisit Chapter 1.)

We take the example of an organization that conceives of and manufactures mechanical arms for war amputees.[3] They work closely with clients in order to produce arms that are well adapted, functional, and somewhat elegant. In such cases, managers cannot fully contemplate projects without first considering their clients' input. Project managers and clients must work together to achieve the expected outcome. Helping cement the relationship are prior interactions between project managers and clients; they foster trust, providing these interactions are positive.[4]

---

[1] Srivastava, Shervani, and Fahey (1998).
[2] Griffin and Hauser (1992).
[3] Toledo et al. (2009).
[4] Hadjikhan (1996).

Crucial to the success of a project are the relationships between project managers and clients, and the way marketing experts implement their strategies. We encourage readers to take their time and read the verbatim and information included in the tables presented in this chapter, as they provide a good feel for how stakeholders experience and live their projects. We take pride in the fact we corroborate theory with practical examples extracted from the literature and from our own research (see Chapter 6, for example).

CHAPTER 5, CLASS EXERCISE #1:

Read the various tables included in this chapter and relate them to personal experience.

PROPOSED QUESTIONS FOR DEBATE:
1. Is the expression "The client is king" fair?
2. Are project managers always right?

Companies can use customers as a source of innovation by asking what they think or experience (e.g., in focus groups or with on-site questionnaires, as done in many hotels), accessing their blogs, listening to their complaints or suggestions, and observing how they use the products. Experts sometimes refer to **conjoint analysis** when product developers work in tandem with clients to develop new concepts.

The learning objectives of this chapter are to understand the relationships between marketing and project management stakeholders, and the importance of including them and clients/end users, and to identify sensitive areas where they must work closely together to ensure success.

A sound understanding of building customer relationships through psychology is fast becoming a key skill in project management.

5.2 Project Managers and Clients

The differences between marketing and project managers stem from two facts: The former focuses on ensuring customers' long-term loyalty (potentially a lifetime) by maximizing benefits and minimizing sacrifices, while the latter is interested in performance over the term of the project (less than five years for many) and reaching the launch date within cost and in line with quality requirements.

Marketing efforts within projects develop according to four phases (Figure 5.1):

1. Pre-project marketing,
2. Marketing at the start of a project,

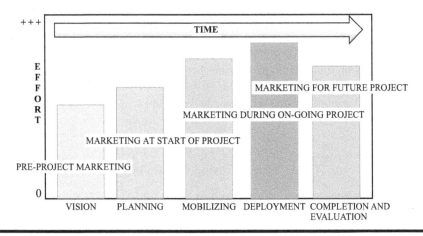

Figure 5.1   The four phases of marketing efforts.

Note: Each phase involves a different emphasis and marketing effort.

3. Marketing during project mobilization and deployment (so-called ongoing projects), and
4. Marketing "intended to create the conditions of a future project."[5]

Indeed, marketing efforts that embrace the entire lifecycle of a project are better at gaining end users' interest.

CHAPTER 5, CLASS EXERCISE #2:

For a project of your choice, and using the knowledge you have acquired thus far, explain how you would go about the four phases of marketing efforts.

5.2.1 Clients' Needs and Involvement

Some projects are more heavily geared toward clients' needs than others. Recovery projects—dedicated to rebuilding infrastructures after natural or human-generated disasters occur (currently mushrooming worldwide)—highlight the intricacies of a close symbiosis. To that effect, the Project Management Institute's (PMI) website,[6] as well as various authors, recognizes the importance of interactions between clients and project managers.[7]

---

[5] Lecoeuvre-Soudain, Deshayes, and Tikkanen (2009, p. 37).
[6] See www.pmi.org/, accessed November 2017. PMI delivers the PMP (project management professional) accreditation.
[7] See, for example. Tikkanen, Kujala, and Artto (2007).

Table 5.1 Some Determinants of CRM

Determinants of Sound CRM	Underlying Construct	4Ps[a]
Psychological Aspect		
"Understanding main customers"	Self (identity)	People
"Consideration of main customers' trends"	Self	
Determinants of Sound CRM	*Parameter*	*4Ps*
Project Triangle (Calendar, Costs, Norms of Quality)		
"Discussion of functionality and usability [resulting from] user involvement"	Quality	Process

[a] The 4Ps of project management: plan, processes, people, and power.

Indeed, experts have identified the importance of consulting with clients and seeking their acceptance as a key success factor (KSF).[8] In fact, clients may be involved in a project in many ways. Some scholars mention five of them: self-service, experience, self-selection, design, and emotional.[9] Table 5.1 uses the marketing concept of customer relationship management (CRM) to provide some examples of what project managers have to say about clients' participation in projects.[10] We separated the answers, given in verbatim, into two categories: psychological and according to the iron triangle of project management. We also identified the underlying psychological constructs and linked them to the 4Ps of project management. (We will use this approach for the balance of this chapter.)

Marketing science highlights the relationship between a project manager and the clients. It positions the project under a new light, one where psychological constructs also play a key role. Of note, however, is that the marketing of *projects* differs from the marketing of a *product* on many fronts,[11] as shown in Table 5.2.

As illustrated by Table 5.2, there are vast differences between the marketing of projects and the marketing of products, mainly because projects are not finished products until the delivery date. Nevertheless, a few psychological constructs permeate either closely or remotely the project environment, namely, customer's satisfaction, involvement/commitment, loyalty (both concepts used mostly in marketing), and cooperation.

---

[8] Pinto and Prescott (1988).
[9] Payne, Storbacka, and Frow (2008, p. 84).
[10] Inspired by Papadopoulos et al. (2012).
[11] Inspired by Cova and Cova (2002), Cova, Ghauri, and Salle (2002).

Table 5.2 The Marketing of Projects Differs from the Marketing of Products

Characteristics of the Marketing of Projects Versus the Marketing of Products	Inferred Psychological Construct	4Ps
In the case of projects, the final product is not 100% known, leading to market-reaction uncertainty.	—	Plan
No substitution available in the case of projects.		
Unforeseen and changing interests of stakeholders when working on a project.		
One can measure customer satisfaction according to front-end experience, through the experience and the end experience.	Satisfaction	Process
No brand recognition for most projects.		
No recall possible; a project is final.		
Ongoing process from beginning to end.		
Product development, product creation, and product placement in the market are unique to each project.		
Key stakeholders are constrained by calendar, costs, and norms of quality.	Involvement/ cognitive dissonance	People
Client's involvement is linked to quality.		
Multi-stakeholder involvement (e.g., public-private partnerships and government projects, causing market competition[a]).		
No phenomenon of cognitive dissonance (post-purchase) and of loyalty because a project is a development effort, not a finished product.		
Customers may provide input as the project evolves (close, active participation or co-construction/cooperation).		

[a] Patanakul et al., 2016.

As mentioned, many projects are hybrid in nature, in the sense they offer both a product and a service at the same time.[12] Thus, they inherently have a psychological connotation. (Service generally implies human interaction.) Infrastructure, such as a highway, is both a product and a service to the end users.[13] In Canada, the Confederation Bridge linking the provinces of New Brunswick and Prince-Edward Island is a marvelous product that also serves the many tourists who cross it each summer. It is so convenient they accept the C$46 price tag (roughly USD 35). In that sense, the marketing of projects is a precursor to the marketing of the product the project will realize within the confines of the calendar, costs, and norms of quality. The customers estimate the value of a project (and hence their satisfaction) through respecting the preset norms of quality tied to design and functionality, costs, and timely delivery.

As seen in Chapter 3, one key characteristic of projects is that they are a plan for the future—a promise of sorts. There exists, however, a potential for overpromising.[14] This leads to many of the problems encountered in projects worldwide. Projects are a collection of various initiatives[15] that individual managers bring to completion in order to make the entire sum of all initiatives efficient. Contrary to any product readily available for consumers, projects are a process and, as such, the client's perspective differs. Project managers *live* the project rather than *consume* a product. A project contains provisions for the future, such as routines.[16] One example of a provision is the replacement of the Champlain Bridge, located in Montréal, Canada, which we mentioned before.[17] This project commanded a prefeasibility study, using simulation, in order to evaluate future travel flows and patterns according to various scenarios. Indeed, projects demand high degrees of foresight.[18] Clients who participate in the realization of a project are active both at the start and later on, so their psychological implication spreads over time and not merely at one particular point of time.

This is why massive projects often take several years to complete. As an extraordinary example, in Barcelona, the Sagrada Familia church remains incomplete fifty years after its construction began. During such a long time span, stakeholders and priorities change, as do the political environment and/or strategies.[19] Under these circumstances, excessive "lifecycle costs"[20] dislodge the project from its normal track.

---

12 Recall that, once realized, the project becomes an operation.
13 Maloney (2002).
14 Gutierrez and Friedman (2005).
15 Williams, Ferdinand, and Pasian (2015).
16 Stephens and Carmeli (2016).
17 See JCCBI (2011).
18 Sanderson (2012).
19 Aritua, Smith, and Bower (2009).
20 Tysseland (2008).

In summary, the marketing of projects certainly uses some marketing techniques and resorts to some marketing theories, but it also has its own peculiarities. From early on, clients may engage in a project yet the perception of the said project—and the psychological frameworks between present, or future, clients and the project managers—may differ. Psychological constructs play a key role at the mutual frontier of marketing and project management.

5.2.2 How Do Clients and Project Managers Consider Each Other?

Project managers may benefit from including clients in different project phases. More particularly, clients experience the iron triangle of project management (calendar, costs, and norms of quality) in two ways:

1. By being part of the project and/or
2. By being the end user of the product a project has created.

This does not mean, however, that efforts at co-development are problem-free. As seen, this certainly points to the underlying presence of psychological constructs.

One such construct is trust; Table 4.18 seen earlier can be read as a barometer of trust as found in the literature.[21] Ideally, marketing and project managers, as well as clients, position themselves within a zone of comfort without falling for blind trust. (Vigilance, in project management, is mandatory.)

Assuming they are within their zone of comfort, participating consumers engage, generate, or foster dialog with project managers.[22] They will also engage with the corporations that have hired these managers (corporations, however, remain faceless). Indeed, increasingly, clients want to participate in projects and project managers want to involve well-chosen customers. This process is one by which "the supplier [in this case, the project manager] supports the customer's practices with an extended offering, including goods components and a range of service activities, which enable the customer to create value out of the core process [e.g., a production process]."[23] Clients expect managers will understand, empathize with, and take into consideration their viewpoints, show respect, and integrate them further in the full realization of projects. As such, they want to rest assured there is accountability and that project managers and their organizations make decisions

---

[21] Mesly (2011).

[22] Facial expressions are universal, allowing different cultures to exchange in a form of communication (Ekman, 1999; Hansen and Hansen, 1988).

[23] Grönroos (2011, p. 241).

in the best interest of the project's end users.[24] Clients seek support from project managers; in turn, support from top management is a necessary ingredient for the project managers' drive to succeed.[25]

Along those lines, NASA's concept of earned value management (EVM) notes, "people who are familiar with processes are needed to ensure adherence to discipline and provide a decision-making point. As a result of good decisions, better customer relations can be attained…"[26] Clearly, project managers and clients engage in co-evolution, which implies they interact and engage with their psychological selves.

The fact that clients expect project managers to understand their needs and maintain good communication may be, in part, what segregates success from failure. Yet, project planners must also be sensitive to the financial capabilities of their clients. Apart from the complexity, scale, and scope of a project, as well as market conditions, this consideration adds to other factors that enter into the formulation of a project bid.[27]

Clients may adopt one of three forms of participation, each with its own psychological flavor (behavior, needs, and commitment). These are "design for," "design with," and "design by," defined as follows:

> Design for: denotes a product development approach where products are designed on behalf of the customers. Data on users, general theories and models of customer behavior are used as a knowledge base for design. This approach often also includes specific studies of customers, such as interviews or focus groups.
>
> Design with: denotes a product development approach, focusing on the customer, utilizing data on customer preferences, needs and requirements as in a 'design for' approach, but, in addition, includes display of different solutions/concepts for the customers, so the customers can react to different design solutions.
>
> Design by: denotes a product development approach where customers are actively involved and partake in the design of their own product.[28]

Certainly, when justified to do so, clients feel they can bring value through a process of collaboration with project managers. Yet project managers also have expectations toward clients they include in their projects. Both marketing and project managers share one interest in particular: They are, or should be, after what marketing experts label as customer lifetime value (CLV), defined as "the

---

[24] See Bruelious, Flyvbjerg, and Rothengatter (1998).

[25] Ratcheva (2009).

[26] Cooper (1979, p. 95).

[27] As well as other factors, for example, location in the construction sector (Akintoye, 2000).

[28] Kaulio (1998, p. 21).

present value of all future profits obtained from a customer over the life of his relationship with a firm."[29] Obviously, CLV diminishes with the emergence of problems.

However, different organizations view clients from contrasting angles. When clients participate in the co-development of a project, managers see them as partners, part-time workers, resources or, in the best of cases, as a way of developing a sustainable organizational advantage.[30] Under ideal circumstances, marketing experts and their sales force know that information about clients' motivations promotes the success of a project.[31] Experts have noted this improves new product development (a project by itself) when paired with research and development (R&D) efforts.[32] When projects go astray, however and sadly enough, managers may be tempted to consider clients as a nuisance in an attempt to avoid accountability.[33, 34]

5.2.3 Benefits and Problems to Clients' Involvement

Practitioners and academics have long recognized the value of getting early involvement from clients.[35] Through complaints, for example, customers ultimately bring value to a firm or a project, as they point toward areas of improvement. These experts warn, however, about the wrong type of client involvement:

> Listen carefully to what your customers want and then respond with new products that meet or exceed their needs. That mantra has dominated many a business and it has undoubtedly led to great products and has even shaped entire industries. But slavishly obeying that conventional wisdom can also threaten a company's ability to compete.
>
> The difficulty is that fully understanding customers' needs is often a costly and inexact process. Even when customers know precisely what they want, they often cannot transfer that information to manufacturers clearly or completely.[36]

Table 5.3 lists some benefits and costs associated with co-evolution between project managers and clients.[37] (We will review these benefits in more detail in the next section.)

---

[29] Gupta and Zeithaml (2006, p. 724).
[30] Hillman and Keim (2001).
[31] See, for example, a study done by Moenaert et al. (1994) on some 40 Belgian firms.
[32] Ernst, Hoyer, and Rübsaamen (2010).
[33] Brady and Davies (2010).
[34] Akintoye and MacLeod (1997, p. 35).
[35] Gruner and Homburg (2000), Brockhoff (2003).
[36] Thomke and von Hippel (2002, p. 74).
[37] Inspired by Ahola et al. (2008).

Table 5.3   Benefits and Costs

	Benefits	Costs	4Ps	Psychological Constructs
Short term	Better product performance, reliability, product, consistency, quality, customization	Conflicts, negotiations	Quality/ processes	Pride/conflict
	Agility	—	Process	—
Long term	Improved image, trust, solidarity, mutual goals	Opportunism		Apprehension/ trust, self, collaboration

Rule of Thumb: Project and marketing managers want to ensure the benefits of involving clients outweigh the drawbacks.

5.2.4 Benefits

There are numerous benefits to developing projects that encourage input (and involvement) from existing clients and/or future customers.[38] Indeed, the intensity of the involvement and efforts granted by clients correlate positively with project success.[39] A meta-analysis of the literature addressing clients' inclusion in projects leads to the conclusion that their involvement in the vision and deployment phases correlates significantly with speed to market.[40] Overall, cooperative efforts

1. May foster benefits provided by other assets;[41]
2. May help generate novel ideas or better processes,[42] which, when applied to design, can positively impact sales (see, for example, in the car industry);[43]
3. May increase the chance of success,[44] or diminish the possibility of failure;

---

[38] Tikkanen, Kujala, and Artto (2007).
[39] Voss (2012).
[40] Chang and Taylor (2016).
[41] Homburg, Schwemmle, and Kuehnl (2015).
[42] Ritter and Gemünden (2003).
[43] Landwehr, Wentzel, and Herrmann (2013).
[44] Muller and Jugdev (2012).

4. Offer project managers a source of knowledge and data[45] that they can use to create value for customers; and
5. Provide project leaders with an opportunity to consider potential future business.

To further this last point, as would-be end users of a project, clients may

1. Have technical or intuitive knowledge about a project's requirements,
2. Have some other expertise to share,[46]
3. Participate in trials or focus groups,
4. Provide feedback once the project becomes an operation (i.e., once the project is complete and used on a regular basis by its intended customers),
5. Serve as leverage toward other potential end users, or
6. Verify that the project meets requirements (in turn associated with needs).

The best clients to integrate into project's processes are those who have operational, technical, and/or leadership skills (as well as stable personality traits!). In the same vein, academics have found a significant correlation between maturity, competence (associated with trust), and support within the management group, and project success.[47]

Yet the benefits are not one-sided; customers also enjoy them. They have confidence that the project managers properly address their needs, that their participation during the transformation phase adds value to the project, and that their collaboration with project managers enhances the overall project experience.[48]

Table 5.4 gives an overview of the benefits of clients' participation in a project and highlights the role of clients' needs, that is, of their motivation. Table 5.5 provides further findings as to the motivation to work together with project managers.[49]

As illustrated in Table 5.5, there are a number of valid reasons for a project's team to engage clients in some form or another, with many linked to the construct of trust. These varied reasons tend to fall under two categories:

1. To reduce problems and
2. To improve the overall project.

---

[45] Cherns and Bryant (1984).
[46] See Jónsson (2012) on the role of outside consultants who bring a new perspective to a project. In many instances, clients can serve as outside consultants of sorts.
[47] Besner and Hobbs (2013).
[48] Zhai, Xin, and Cheng (2009).
[49] Inspired by Lu and Yan (2007).

Table 5.4 Benefits to Client's Participation in Projects

Benefit	Inferred Psychological Construct	4Ps
In dynamic and complex environments, clients may help focus the efforts of the team.	Motivation (needs)	Plan
One can regroup multiple goals, interests, and strategic considerations according to clients' needs, as expressed by them.		
One may devise provision for the future to fit clients' actual needs, as expressed by them.		
Interdependencies among projects may be more visible when viewing them from a client's perspective.	Needs	Process
Uncertain or changing information may be alleviated when clients properly formulate their needs.		
The presence of many decision-makers[a] spread over various locations (especially in international projects) commands a highly client-focused approach.[b]	—	People

[a] Haverila and Fehr, 2016.
[b] Ringuest and Graves, 1999.

Such inclusion provides positive enhancement to a project only if the relationships between project managers and clients are harmonious. Clients' requests for rework, for example, can significantly delay a project or annoy project managers if they find these requests unjustified. Table 5.6 delves further into a client's willingness to cooperate.[50]

As can be seen from Table 5.6, building mutual trust leads to healthy cooperation.

---

[50] Inspired by Crespin-Mazet and Ghauri (2007), pp. 162–163.

Table 5.5 Possible Reasons to Engage Would-Be Customers

Reasons to Engage	Inferred Psychological Construct	4Ps
To "reduce engineering rework."	Conflict	Process
To "reduce litigation."		
To "enhance company culture."	Trust motivation (needs)	People
To "enhance reputation" (related to trust).		
To "establish long-term relationships."		
To "improve long-term competitive advantages."		
To "improve social responsibilities."		
To "obtain the support of partner's expertise and knowledge."		
To "penetrate new markets."		
To "serve core customers" (answering clients' needs).		

Table 5.6 A Client's Willingness to Cooperate

Client's Motivation to Cooperate	Underlying Psychological Construct	4Ps
	Will most likely participate if…	
"The customer's behavior toward co-development is dependent on their perception of the risk associated with the project transaction."	Apprehension/risks are low	Plan
"The higher the perceived risk for the customer, the more positive the attitude toward co-development."[a]		
"Customer's business culture: transactional *versus* relational orientation."	Collaboration is high	Process
"Customer's purchasing routines and strategy."	Control (sense thereof) is high	

(*Continued*)

Table 5.6 (*Continued*)   A Client's Willingness to Cooperate

Client's Motivation to Cooperate	Underlying Psychological Construct	4Ps
"The customer's behavior toward co-development is dependent on their perception of the risk associated with the relationship to the supplier and with the network actors."	Apprehension/risks are low	People
"Shared values and expectations."	Collaboration is high	
"Desire of the parties to develop the necessary resources, efforts, and investments to create a successful relationship."	Commitment is high	
"Proximity/commonality of goals."	Distance is low	
"Co-development requires trust between parties."	Trust is high	
"Existence of shared goals and values."		
"Customers are inclined to co-develop the project with the contractor when their level of knowledge is low; they lack competences or availability and they lack financial resources."	Trust is high	
"The customer is more willing to co-develop the project with the contractor if they have identified that the contractor has a specialized knowledge they do not have."		
"Project network characteristics."	Trust/collaboration is high	
"Trust is required to engage in co-development as there is a strong dependency between the actors."	Trust/dependency is high	

[a] *Note:* Where we saw fit, we corrected the English used in the original text, to clarify the meaning of the sentences.

CHAPTER 5, CLASS EXERCISE #3:

Choose a project for which you are the project manager. List, and briefly explain, why you would want customers' participation.

For the same project, list and briefly explain why you would not want customers' participation.

PROPOSED QUESTIONS FOR DEBATE:

1. There is a disagreement between the project manager and the marketing manager. Who should have the final word?
2. There is a disagreement between the client and the marketing manager. Who, among all stakeholders, has the final word?

5.2.5 Failures and Problems

Failures have traditionally been associated with cost overruns, delays, quality problems, or a mixture of these factors. As mentioned previously, clients are unlikely to be blamed for failures—an observation that emphasizes the project manager's role.[51]

On occasion, however, tugs-of-war occur between project managers' and clients' interests and power, especially with respect to their capacity to dictate the desired course of action. One major inconvenience faced by project managers due to the involvement of multiple stakeholders is the prevalence of unanticipated, or else unforeseen, changing interests.[52] In addition, client involvement may cause costs to swell.[53] Academics posit that the maturity and the dedication of both top management and project managers have a positive effect on the three main measures of a project (iron triangle): calendar of tasks and activities, costs, and norms of quality.[54]

In general, many problems are due to either shortages or excesses, something we saw in our discussion of point of equilibrium (POE) in Chapter 3. Over-optimism in the planning process (making unrealistic projections) is a behavior that transpires from the psychological construct of self. Under-evaluation of a project's challenges, labor productivity, and costs, as well as of its capacity in terms of control, standardization, and supply management, will lead to shortages. Table 5.7 gives an

---

[51] Dvir et al. (2006).
[52] Verweij (2015).
[53] Peled and Dvir (2012).
[54] Berssaneti and Carvalho (2015).

Table 5.7 Examples of Problems with Their Psychological Dimension

Example	*Problem*	*Inferred Psychological Construct*
Microsoft Xbox360	"Groupthink was also an issue."	Collaboration
Maritime Coast Guard Domain Awareness Project	"Project leaders succumbed to the illusion of control bias."	Control
Columbia Space Shuttle	"It was a culture where people were unwilling to speak up."	Control/ collaboration
Mars Climate Orbiter/Mars Polar Lander[a]	"Selective perception [...] design team failed to coordinate with the operational team at NASA."	Perception/ collaboration
Denver Airport baggage handling	"Overconfidence [... and] the sunk cost"; estimated cost overrun.	Psychological self and structural (costs)
New York City subway communications system	"Conservatism, overconfidence, and illusion of control."	Self/perception/ control
Example	*Problem*	*Structural/ Mechanical*
Denver International Airport	"Exceeded the original budget by 200% and was delivered sixteen months over schedule."[b]	Budget
Eurotunnel	"Average 69% increase in costs from 1986 vs. 1994."[c]	Costs
FIFA World Cup 2014 project budget increased	"Originally estimated at €1 billion; revised to €11 billion."	
Merck Vioxx	"Sunk cost, trap dominated."	

(Continued)

Table 5.7 (*Continued*)   Examples of Problems with Their Psychological Dimension

Example	Problem	Structural/ Mechanical
Airbus A380	"Failed to transform itself from a balkanized organization into an integrated company."	Process
Heathrow Airport, Terminal 5 (finished on time and within budget, but with a number of problems at launch)[d]	"(1) Troubles locating parked cars, (2) signage problems, (3) baggage handling delays, (4) unserviceable escalators, (5) electrical problems, (6) guidance systems incorrectly calibrated, (7) failure of the automated temperature controls."	Planning/quality

[a] Sauser, Reilly, and Shenhar (2009).
[b] Flyvbjerg (2005).
[c] Winch (2013).
[d] Brady and Davies (2010).

example of such problems. Again, we extract from the main source[55] the text that best highlights, in our opinion, the underlying psychological constructs at play.[56]

Table 5.8 continues[57] along the lines of Table 5.7.

5.3 What Are the Nine Factors of Trouble?

Based on a considerable review of the scholarly literature, we created a succinct list of nine factors that may cause relationships between project managers and clients to derail. The marketing feasibility analyst should scan the environment of the project to check for their presence and prevalence, especially in the case of existing project groups that contain members (and possibly clients) who have worked together in the past. The nine factors are as follows:

1. Pressure to perform,
2. Rework,
3. "Demotivating" factors,

---

[55] Shore (2008).
[56] Inspired in part by Shore (2008).
[57] Inspired by Ruuska et al. (2009).

Table 5.8 Classifying Problems

Problem	Underlying Psychological Constructs	4Ps
"Potential hidden agendas."	Apprehension	Plan
"Incomplete quotation information."	Collaboration	
"Misaligned objectives."	Control	
"Unclear roles and responsibilities."		
"Inappropriate selection principles for suppliers and contractors."		
"Inappropriate contract types and adherence to contracts."	Trust/ collaboration	
"Inadequate documentation procedures."	—	Process
"Firms' incomplete systems and processes."	Collaboration	
"Insufficient communication structures and mismatch between communication purpose and style."		
"Action or inaction based on assumptions, rather than facts."	Commitment	
"Lack of experience and capabilities."	Trust	
"Lack of knowledge of specific [local] requirements."		
"Diversity of actors."	Distance	People
"Lack of trust."	Trust	
"No previous joint-working experience."		

Note: These constructs are some of the most commonly inferred in literature on project management, human resources, and marketing, although they are not always explicitly treated by their names.

4. Discretion,
5. Escalation of commitment (or over-commitment) *versus* loyalty,
6. Entrapment,
7. Client inadequacy,
8. Diverging interests, and
9. Particular problems in less-advanced countries.

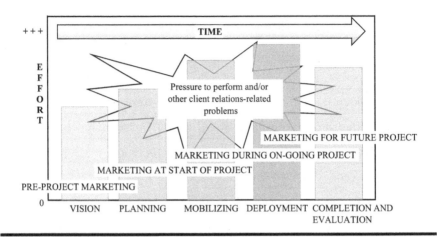

Figure 5.2   The four phases of marketing efforts, revisited.

Note: Here, we show how the nine factors of trouble (identified through our extensive literature review) affect marketing efforts.

We examine them in the sequence as shown in Figure 5.2.

5.3.1 Pressure to Perform

The pressure to perform has two sides: For one, clients may be a motivating factor to complete the project on time,[58] yet they may also become a hindrance if they push their own agenda too strongly.

Projects, nowadays, are under considerable pressure to bring the products they develop to market. This comes from a variety of influences, including[59]

1. Competition;
2. Political agendas;
3. Public opinion,[60] and social groups such as Greenpeace; and
4. Timeline[61] (indeed, many projects, especially complex ones, are developed over several years).[62]

This leads us to infer that the environment, and the people within and outside the project's organization, is likely to change, causing additional uncertainty. As

---

[58] Alderman and Ivory (2007).
[59] Vicente-Oliva, Martínez-Sánchez, and Berges-Muro (2015).
[60] Chih and Zwikael (2015).
[61] Bryde and Robinson (2005).
[62] Patanakul et al. (2016).

pointed out by some authors, to "remain effectively related to rapidly changing environments, firms are periodically faced with the challenge of re-deploying their existing resources and changing their internal processes and capabilities."[63] Not only are project environments fast-paced, but they are also increasingly multicultural. These cultural differences may cause distrust from stereotypes and can put pressure on the capacity to perform.[64]

5.3.2 Rework

Clients may occasionally demand rework. A study investigating 381 projects performed by 51 companies showed that 80% of companies experience client-related rework, causing, on average, almost three and a half weeks of delays.[65] We listed the main causes of rework requests in Table 5.9 and added our interpretation of the nature of the problem with respect to mechanical/structural or psychological constructs.[66]

As Table 5.9 shows, we have associated many rework demands with a psychological component. This again illustrates that the human aspect of the relationship between project managers and clients is instrumental.

Table 5.9 Client Demands for Rework

Demand	Nature of Problem
"Inadequate/incomprehensive project objectives by the client."	Psychological/cognitive
"Financial problems faced by the client."	Cost
"Change of plans or scope by the client."	Various causes
"Obstinate nature of the client."	Psychological
"Impediment in prompting the decision-making of the client."	Psychological/social
"Replacement of materials by the client."	Technical
"Change in specification by the client."	

Note: "Obstinate nature" refers to a judgment of character or personality.

---

[63] Davies and Brady (2000), p. 934.
[64] See van Marrewijk and Smits (2016).
[65] Hwang, Zhao, and Goh (2014).
[66] Inspired by Hwang, Zhao, and Gohm (2014), p. 704.

5.3.3 Demotivating Factors

Clients' constant demands may alienate managers from the project, if not discourage them from completing it. Apprehension between these two stakeholders may build up,[67] and a number of other factors may discourage project managers. In Table 5.10, we provide possible causes of "demotivating factors" associated with key project constructs.[68] The readers can observe some causes of lack of motivation are due directly, or indirectly, to clients (indicated by *). According to some

Table 5.10 Demotivating Factors

Demotivating Factors	Rank[a]	Underlying Psychological Constructs	4Ps
"Inadequate salaries and rewards."	6	Fairness	Process
"The working environment is focused on negative criticism."	37	Self	
"Perceptions of lack of respect among coworkers."	41	Apprehension	People
"Uncooperative behaviors of clients/ other project stakeholders"*; "poor communication within design teams."	15, 5	Collaboration	
"Inadequate commitment among design team members."	21	Commitment	
"Incompatibility of design team members"; "unhealthy competition among coworkers."	12, 14	Conflict	
"Distrust and dishonesty among design team members."	38	Trust	
"Inept leadership, personal behaviors (e.g., dictatorial, bullying, intimidation)."	8	Control	Power
"Lack of open interaction between leaders and subordinates"; "poor feedback and inappropriate evaluation system."	17, 19	Transparency	

[a] Out of 43 demotivating factors identified by the authors.

---

[67] Chen (2011).
[68] Inspired by Oyedele (2013).

academics,[69] examples of sources of project-induced stress include[70] poor relationships (75%), excess paperwork (69%), pressures due to time (67%), staff inadequacy (43%–67%), conflicts (50%), and work overload (44%).

A study of 95 U.S. organizations found that a poor organizational culture[71]

Project's Triangle of Constraints		
Demotivating Factors	*Rank*	*Parameter*
"Working excessively long hours."	9	Time
"Design decisions are dictated by cost and not quality factors."	3	Cost
"Clients demand for radical for high degree of innovation in design"; "frequent design changes/variations."	2, 3	Quality

Note: Trust is one of the major psychological constructs evoked in the context of relationships.

accounts for a very high percentage of failures in product development, as well as a substantial percentage in product launch difficulties.[72] This goes to show that internal work psychodynamics[73] (PWP) of a particular project may indeed affect its course. This example emphasizes the importance of sound relationships, or of what some academics call "hygienic relationships," between stakeholders.[74] While it is not the project managers' job to motivate the clients, it certainly falls within their responsibility to maintain hygienic relationships or, put differently, to have "good interpersonal skills, [and the] capability to build mutual trust and cooperation."[75]

A Swedish study of 15 contractors leads its author to stipulate that standardized processes do not impede motivation.[76] We posit that perhaps the reverse is also true: Well-designed plans and well thought of work tasks encourage people to perform

---

[69] Sutherland and Davidson (1989).

[70] In parentheses are the source percentage levels according to a survey of 71 construction project managers.

[71] Recall we defined the project culture by the 4Ps of project management: plan, processes, people, and power.

[72] Belassi, Kondra, and Tukel (2007).

[73] Typically, PWP include control, transparency, trust, fairness, collaboration, and commitment (Mesly, 2015).

[74] Hygienic relationships can also be referred to as an "interfunctional climate" (see Voss, 2012, p. 571).

[75] Peled and Dvir (2012, p. 325).

[76] Polesie (2013).

Table 5.11 Constructs Identified in a Chinese-Led Project

Psychological Aspect		
So-Called Social Risks	*Demotivating Factors (Rank in Brackets)[a]*	*Underlying Psychological Construct*
"Social conflicts"	"Poor communication within design teams." (5) "Poor coordination of design teams." (11) "Uncooperative behaviors of clients/ other project stakeholders." (15) "Counterproductive behaviors from coworkers." (23) "Distrust and dishonesty among design team members." (38)	Conflicts
"Unfair land compensation"; "unfair house demolition compensation."	"Inadequate salaries and rewards." (6) "Inadequate opportunity for career development/promotion." (14)	Fairness

[a] Ranked out of 43 sources of motivation, as listed by Oyedele (2013).

because they have a sense of direction and a sense of security that their efforts will not be lost.

Table 5.11 highlights some of the points of our discussion, with excerpts from a text that describes a Chinese-led project.[77]

Project's Triangle of Constraints	
"Environmental pollution"; "concerns on the standard of living."	Quality

Note: Conflicts are one of the major constructs evoked by project managers in general.

Adding this Table 5.11 to other tables we have examined, we can safely claim that problems know no cultural frontiers.

---

[77] Inspired by both Shi et al. (2015) and Oyedele (2013).

5.3.4 Discretion

The secrecy of certain parts of a project may require a level of discretion that clients may not be able to sustain.[78] Some prime examples include the Manhattan Project or Apple's product development strategy.

5.3.5 Escalation of Commitment (or Over-Commitment) Versus Loyalty

Escalation equates to a need to justify or, according to the tenet of project management theory, comply with self-congruence.[79] Academics have long studied commitment escalation, including the case of a disastrous climb of Mount Everest. Stakeholders estimate that, past a point-of-no-return (a concept we saw in Chapter 3), it is cheaper to continue with the project than to stop it, even if failure entails perilous outcomes.[80] Many other examples exist,[81] including the Expos '86,[82] the Shoreham Nuclear Power Plant, Chicago's Tunnel and Reservoir Plan (the Deep Tunnel Project), and the Ravensthorpe Nickel Mine.

Some authors argue that the future of doomed projects is dependent upon commitment escalation, the strategic misrepresentation of the very future of the project, as well as to technical errors generated by over-optimism.[83] Other authors add the project itself and organizational variables as key determinants,[84] citing four factors: biased information, self-justification, self-reinforcement, and social pressures. We can link the last three to psychological constructs, such as emotional biases. Table 5.12 gives a portrait of escalation determinants as we see them.[85]

5.3.6 Entrapment

A term akin to escalation is that of entrapment.[86] Some academics have defined entrapment by discussing the following four criteria:

1. Absence of ergodicity (together considered to be external),
2. Absence of predictability or risk,
3. Inefficiency,[87] and
4. Lack of flexibility in change management.

---

[78] Chang and Taylor (2016).
[79] Brockner (1992).
[80] Alvarez, Pustina, and Hällgren (2011).
[81] See, for example, Meyer (2014).
[82] Ross and Staw (1986).
[83] Winch (2013).
[84] Ross and Staw (1986).
[85] Inspired by Meyer (2014).
[86] Brockner et al. (1984).
[87] Sydow, Schreyögg, and Koch (2009).

Table 5.12 Escalation Determinants

Escalation Determinants	*Underlying Psychological Construct*	*4Ps*
"Norms for consistency."	Self	Process
"Saving face." "Self-justification."		People
"Optimism and illusion of control."	Control	Power
"Invulnerability effect."		
Project's Triangle of Constraints		
Escalation Determinants	*Parameter*	*4Ps*
"Sunk cost effect," whereby it appears cheaper to continue than to cross out the investment done so far.	Costs	Plan
"Government intervention."	Control	Power
"Legal implications."	Costs	

Note: Self (self-image, ego) and sense of control are psychological constructs that contrast with the "mechanical" aspect of management entailed in cost control, one of the three elements of the iron triangle of project management.

We consider the latter two to be internal, thus pointing to points of vulnerability (POVs). Other researchers have shown entrapment was high when people feel they are losing their "bet" (their project), especially when the degradation of the project is slow, when the outcome is not clearly laid out, or when there is a possibility of access to additional resources.[88] (In this case, the psychological component is unclear and may be circumstantial.)

5.3.7 Clients' Inadequacy

Some project managers may think it is not worth including clients in the development process of a project. Some companies prefer not to deal with certain clients because they cause them grief and undue costs, while some clients may not have the skills or knowledge needed to participate in specific projects; they may be short on social skills (e.g., diplomacy) or technical skills.[89] Ill-defined needs or unrealistic demands may also weigh on a project's development from inception. On the other hand, some authors[90] discuss the fact that clients who are unhappy with a particular product may actually end up manifesting various

---

[88] Rubin and Brockner (1975).
[89] Oyedele (2013).
[90] Kähr et al. (2016).

forms of instrumental hostility, such as sabotage. Nothing precludes the possibility that clients who are included in a specific project, if they become discontent, resort to counterproductive behaviors, which, of course, will harm said project. Managers must carefully choose clients they wish to work with, just as for team members. After all, such clients will become part of the team, even if only on a part-time or restricted basis.

Because clients receive no pay, we posit they feel they deserve treatment that takes into account their voluntary effort. They are unlikely to accept being taken advantage of, or being treated as second-class partners. Yet they may prioritize efforts too much in their favor and may actually be oblivious to the other stakeholders' needs,[91] not realizing projects are a common effort.

Several empirical researches tend to show that managers must use sound communication strategies to mitigate co-production intensity, which generally develops when project managers and clients engage wholeheartedly in projects.[92]

5.3.8 Diverging Interests

Even though clients are brought on board with the hope they will help, that does not mean they necessarily share the same interests as project managers. Clients may be foremost interested in functionality, whereas project managers may be concerned mostly with costs.[93] Indeed, the ratio of benefits to costs differs depending on the stakeholders.[94]

Furthermore, various stakeholders have different approaches to projects depending on such criteria as their perceived power, legitimacy, or urgency. This, in the end, may alter their support and inherent capacity to exert influence.[95]

There may be hidden agendas or contradicting underlying goals.[96] Some authors[97] consider that customers' expectation of costs may actually influence project managers' estimates. At the same time, mismatched expectations make for conflicts and disillusion.[98]

We assume that clients are concerned with satisfaction and project managers with time, costs, and quality. Satisfaction results from respecting the delivery time, costs, and norms of quality; the promise[99] has been fulfilled. Stakeholders may vary

---

[91] Bryde and Robinson (2005).

[92] See the literature review by Haumann et al. (2015).

[93] Yu-Chih and Yuliani (2016).

[94] Andersen, Söderlund, and Vaagaasar (2010).

[95] Aaltonen et al. (2015).

[96] See Ruuska and Teigland (2009).

[97] Jørgensen and Sjøberg (2004).

[98] Gutierrez and Friedman (2005).

[99] Nowadays, with the increasing tendency to buy products on the Internet and get them delivered to a home, the relevance of the promise is even greater.

in their take on the project,[100] however, and this may be a major source of conflict and can impede projects.[101] Perhaps diverging interests are not the actual cause of friction. Often, it is the difference in perceptions, notably with respect to risks[102] and project identity, which unwittingly separates rather than unites project managers and clients.

It may be too hard, too late, or too costly for clients to switch project managers; they are then stuck with them, and the reverse is perhaps true. This leads to a possibility of abuse, or opportunistic behaviors. Knowing that other stakeholders feel stuck with the deal may cause one to take advantage of the other.

5.3.9 Particular Problems in Less-Advanced Countries

Less-advanced countries face particular problems that impede project advancement and completion. They include the lack of an adequate workforce, deficient training and motivation, even from eventual end users, a situation worsens if these users participate fully in the project from beginning to end.[103]

CHAPTER 5, CLASS EXERCISE #4:

Choose a project in which you have been closely involved, even a personal one. Review which of the nine factors of trouble you experienced, and explain their effect on both you and the project.

PROPOSED QUESTIONS FOR DEBATE:

1. Among the nine factors, are some inevitable?
2. As a feasibility analyst, should you expect a project to suffer from multiple utility drawbacks expressed by all or some of the above distressing factors?

5.3.10 Summary of Benefits and Problems

The pressure to perform due to the presence of competition and/or the media, contractual clauses that include penalties for not delivering on time, the tendency to save face, the attempt to rely on past achievements to pursue an obviously flawed venture, and so on may all play a role in doomed projects. All can cause frictions between project managers and clients. Because clients represent both an opportunity and a potential problem, project managers may tend to seek moderate

---

[100] Heravi, Coffey, and Trigunarsyah (2015).
[101] Mesly (2017).
[102] Abednego and Ogunlana (2006).
[103] See, for example, Tabassi and Abu Bakar (2009).

Figure 5.3 A simplified and stylized model of psychological constructs *versus* outcomes.

Note: When psychological constructs are positive (e.g., trust), the influence is positive. When they move toward negativity, the project and marketing outcomes become increasingly geared toward failure.

relationships. Certainly, positive psychological constructs facilitate the emergence of positive outcomes, including technical/mechanical ones, such as technical project success (see Figure 5.3 for an emerging and stylized model).

In Figure 5.3, of course, positive influence is impeded in its functionality by the nine factors discussed above, or at least by some.

5.4 Conclusion

One cannot dissociate projects, project managers, and the clients. Between them, a marketing relationship exists even in the absence of a marketing manager; the project manager becomes in this circumstance the marketing manager of sorts. In many cases, clients are the future owners, or end users, of the projects (deliverables)[104] and, as such, it is advantageous for both parties to cooperate and produce the expected deliverable. Project managers and clients can form one single entity in the realm of seeing projects through from beginning to end.

Through the examples given in this chapter, we have seen that many phenomena related to project management and client relationships revolve around a limited number of psychological constructs (intimately linked to the core competencies), for both project managers and clients.[105] Additionally, both are constrained by the project's iron triangle (calendar, costs, and norms of quality), which forms the basic framework of the discipline of project management. Psychological constructs include apprehension (control and transparency), trust, fairness, cooperation, and commitment. This is also the case in many different cultures (e.g., African, Asian, European, American) we examined throughout this chapter.

A project manager's vectors must align with that of the clients to some degree; compatibility, as opposed to divergence of interests, is crucial for the success of a

---

[104] Kadefors (1995).
[105] Many of which have been found to form core competencies Mesly (2015).

project. The value customer associates with a project benefits from efforts of compatibility. Marketing and project managers endeavor to minimize or eliminate the nine factors we have found that could cause disagreements between them and their clients. As discussed, some authors refer to CLV, which, in terms of projects, would mean that clients would accompany a project from its outset to its turning into an operation—for as long as it lasts.[106] From this perspective, clients themselves are the project. The commonality of interests makes the relationship between project managers and their clients easier to manage and, again, this seems to be true worldwide.

Projects are finite entities; they constitute closed systems[107] where retaliation is possible. Projects require healthy relationships (people) and leadership (power) to develop smoothly.[108]

Clients may be involved from the outset and/or during development, prototyping, testing, trials, and commercialization.[109] One could regard client involvement as a form of investment that brings a return: the success of the project.[110]

The more the deliverable serves its intended clientele (e.g., a bridge or a hospital) over the years, the more clients should be involved from the start. This will give them an opportunity to voice their concerns and desired specifications, and it will help in covering as large a client base as possible. While there are drawbacks in getting clients involved at any point in time on specific projects, the benefits seem to justify that project managers build strong, harmonious relationships with their clients while taking into account psychological constructs such as trust, cooperation, fairness, and commitment.

5.5 Mind Teasers

Readers may use the mind teasers as questions in preparation for an examination or quiz.

1. List, and briefly describe,
   a. the nine factors that may derail relationships between project managers and clients,
   b. at least four benefits of cooperative efforts generated by clients' inclusion in projects, and
   c. at least four positive things clients may bring when cooperating with project managers.

---

[106] Schulze, Skiera, and Wiesel (2012).
[107] See Verweij (2015).
[108] See Mäkilouko (2004).
[109] Coviello and Joseph (2012).
[110] See Smyth and Lecoeuvre (2015).

2. True or False?

a. Clients experience the iron triangle of project management (calendar, costs, and norms of quality) in two ways: by being part of the project and/or by being the end user of the product a project has created.

b. Clients may adopt one of three forms of participation, each with its own psychological flavor (behavior, needs, and commitment): "design for," "design with," and "design by."

c. Conjoint analysis occurs in the context of project management when product developers work in tandem with clients to develop new concepts.

d. Entrapment has been found to depend on four criteria, including absence of predictability or risk and lack of flexibility in change management.

e. Experts have identified the importance of consulting with clients and seeking their acceptance as a KSF.

f. In many cases, clients are the future owners, or end users, of the projects (deliverables).

g. Marketing efforts within projects develop according to four phases: pre-project marketing, marketing at the start of a project, marketing during project mobilization and deployment (so-called ongoing projects), and marketing "intended to create the conditions of a future project."

h. Project teams engage clients in some form or another, with many linked to the construct of trust to reduce problems and to improve the overall project.

i. Projects are a plan for the future—a promise of sorts.

j. Projects, nowadays, are under considerable pressure due to competition, political agendas, public opinion, and social groups such as Greenpeace, but certainly not due to timeline.

k. The marketing of projects differs from the marketing of products.

l. Common psychological constructs found in project management literature include apprehension (control and transparency), trust, fairness, cooperation, and commitment.

Chapter 6

Testimonials

An interior view of a relatively modern building in Nice, France. It was designed to please visitors, and to create a certain atmosphere of self-reflection.

6.1 Introduction

In this chapter, we provide verbatim interviews we have conducted over the years and brief descriptions of the companies we approached. We cover co-construction, core competencies, innovation, the iron triangle, the key consensus factors (KCFs), and the key failure factors (KFFs), and finally, we give examples of actual experiences in product development.

This chapter is meant to read like a snapshot of managerial experiences across a wide variety of businesses. It complements the quotes provided in the numerous tables presented in Chapter 5, in order to illustrate the theory in this book with pragmatic, day-to-day experiences. Learning objectives are to give the readers various portraits of stakeholders and companies involved in a wide variety of projects.

6.2 Testimonials

6.2.1 Co-Construction

6.2.1.1 Michel Dubuc, Aedifica, Architect, Montréal, Canada (2015)

> Our ways of doing things had to be adapted to the harsh reality of the living conditions we faced.

"We arrived in Haiti in 2010 as part of the well-known international reconstruction effort after the devastating earthquake."

"There is no doubt in my mind that there are a number of similarities between the two regions. In both cases, Africa and Haiti, those engaged in implementing projects are forced by circumstances and events to get out of their zone of comfort."

"We discovered that Canadian or U.S. construction regulations, for example, could not necessarily and readily apply in the kinds of environments we are dealing with there. Installing sprinklers, a mandatory device in Canada and often a requirement imposed by international donors, was out of the question for us in Haiti: No one was trained to operate and fix them in case of failure—we had to palliate with other measures, such as devising additional exits."

6.2.1.2 Michel Makiela, Consultant and Professor, France (2019)

> This co-construction turned out to be beneficial to all.

"In 2003, this University decided to develop an MBA continuous education program, with an emphasis on the automotive industry, aimed at young managers in collaboration with major car makers and parts suppliers—an MBA Automotive of sorts. The launch date was set for early September 2005. A team was recruited, and while no one was a specialist in continuous education programs nor an expert in the automotive sector, all members had solid experience in international projects in various sectors such as electronics. The team had, thus, some useful competencies but faced the lack of specific experience related to such an MBA program, which represented a challenge."

"The team first set a committee and launched a targeted media campaign to attract large enterprises, seek the interest of young managers, and identify other stakeholders and local institutions that might be interested, remotely or not, in the project. The team conceived a draft layout of the course load, identified potential professors, devised the possible marketing plan aimed at attracting the target market, and eventually proceeded to select the students and professorial candidates."

"No prefeasibility study was prepared, but the team worked hard to identify the actual needs of the industry, which included project management skills, change management, multicultural management, global supply chain management, and the marketing of innovation."

"Meetings with key stakeholders in the industry revealed what priorities prevailed at the time. The team eventually formed an informal committee composed among other leaders of the industry, who saw the opportunity to form future employees and tailor them to their specific needs, in line with the principle of mutual interest."

The team attempted to validate the offer with all the key stakeholders through numerous meetings and back-and-forth discussions and adjustments. Two concerns were raised at that time:

1. Employees with and without an MBA would compare their salaries and
2. MBA students would seek employment elsewhere, in search for better pay and working conditions.

This led some promoters to think the project was not feasible. The project was henceforth redesigned; training modules would be developed to address very specific needs, especially for young engineers in the automotive industry. Modules would address financial planning, project management, and the like. The program would be a Master of Science, rather than an MBA, would start in September 2006, and would be known as an incubation center, Management Institute Management for the Automotive sector, which would propose not only training but also conferences.

"Potential client fears and objections allowed the promoters to redirect the project to be more in line with the reality of the market."

6.2.1.3 Sybille Persson, Founder of School of Coaching, Nancy, France (2019)

> Work relationships have had more impact on me than any external factors
> I have had to face over the years.

"Business coaching has been an important part of industry development for the past few decades. Managers want to improve their skills, and often need someone with great experience to assist them. While there has been a substantial number of essays and various studies done on coaching, merging theory and real-life experience remain a challenge."

"This state of affairs motivated me to create a school of coaching after discussions with various colleagues and experts, including members of the International Coach Federation (ICF) and with management at the existing Grande École (ICN). Laurent Goldstein, a key partner in the project, has been instrumental from the start and we have always maintained close cooperation."

"Candidates graduate with the title of business coach after successfully following our curriculum. In 2011, our Professional Coach program was the first of its kind to be listed in the French national directory of professional certifications (*Répertoire national des certifications professionnelles*) issued by the Ministry of labor."

"Creating the school, developing it, and managing it have not always been easy. Much to my surprise, problems we faced did not come from outside our small organization and the people we were associated with, but from inside. Furthermore, they related mostly to human factors rather than, say, budget."

"Despite the ups and downs, the school took form and a total of eight people formed the first cohort back in 2008. Finding interested clients/students remained a challenge, one that had us regularly swinging between enthusiasm and worry depending on how things evolved. We realized promotion was crucial, and to raise awareness among the targeted clientele money and marketing efforts had to be invested by all stakeholders associated with managing the emerging school. Although doubts initially hampered budget allocation among some of the key decision-makers, we managed to grow and saw the development of various branches apart from the home base in Nancy. These include Luxembourg (2011) and Paris (2019), with cohorts averaging roughly 15 people."

"Our approach is to encourage innovative pedagogy and a strict path to getting accreditation, including a thesis defense in front of a jury of four experts and the submission of a real case study. It has also been part of our philosophy to celebrate milestones at Nancy's City Hall, something that helps cement the group of stakeholders. Bonding has proved extremely beneficial to the development of the coaching school."

6.2.2 Core Competencies

6.2.2.1 A. Girard, Project Manager, Canada (2015)

It is this combination of efforts and talents that drove us to the finish line.

"Through the course of my career, I have been called to oversee a number of international projects, one of which, as an example, was with the aluminum company Alcan in Arvida, Québec, Canada. The $4.7M CAD project was realized on time, below budget (at $2.96M CAD), and within the quality specifications set up front. What allowed us to successfully complete the project can be attributed to a number of factors. We were inventive; we took great care in defining the project and in addressing environmental concerns right up front. We hired an extremely competent team, which was very much committed and which worked with flexibility and rigor."

6.2.2.2 Brian Saulnier at Seacrest Fisheries, Animal Food Manufacturer, Nova Scotia, Canada (2015)

If I have to summarize our project, it was all about loyalty.

"The project came out of a vision to see equipment and building assembled in a functional manner in order to produce mink food, for which market demand and sales had previously been anticipated. I knew it had to be framed within a budget constraint and a specific timeline.... This is an area where we learned we should have done things differently. Recall we had a business that was mostly based on temporary work and limited skills; the work load evolved with the fishing season. However, with the mink-producing plant, we moved to full-time, year-long jobs. Yet employees never had a job description. Hence, they acquired rights and privileges that at times were counterproductive and that we could not, as managers, change for lack of proper procedures. We are still experiencing difficulties in this area nowadays, albeit they are being addressed... We are a small part of the country. Here, business is done with handshakes, not with lengthy, overly complex contracts. We have close bonds with the employees and our suppliers, and everyone out there seeks to help one another. Management shows gratitude by way of flexibility, for example, with vacation allowance and bonus pay."

6.2.2.3 J. P. Deveau, Algues acadiennes, Sea Products, Nova Scotia, Canada (2016)

> When our team members see growth, they are inclined to jump in and to embrace challenges.

"People like to have a good career and to work for a company where there is a future and a sense of leadership. People like to see a company building its own future. This spawns a lot of self-confidence."

6.2.2.4 G. Oakley, at A.F. Thériault, Boat Building Company, Nova Scotia, Canada (2016)

> Nobody is left out.

"As a project manager, I endeavor to respect knowledge, ideas, and experience team members bring in. Each team member has a voice on each of the ten projects that I head concurrently. Managers lead tradespeople, and tradespeople are assembled according to the project's needs and bonded based on fairness. This is how we work here, and our company keeps growing… We like to make team members part of the decision process. We display a high level of flexibility. In this region of Clare (Nova Scotia, Canada), most of the population is French-speaking; many had been ordered to go to English schools as children, and were at some point in history segregated, and even thrown out of stores for speaking French. Most Anglophones have not had the opportunity to learn French. Yet we all communicate and understand each other despite the language differences. This is the making of a great work team… Team members and managers alike embrace a prosperous future. We encourage ownership of the work; we recognize that everyone, at all levels of the projects and at every step of it, plays an important role. We are part of a team; we are proud. The combination of these positive feelings drives us to outperform ourselves every day; we do our best work day in and day out and look forward to a larger goal. This sense of accomplishment is well represented during milestone events such as the launch at sea of a new vessel, when workers bring in their families and rejoice."

6.2.3 Innovation

6.2.3.1 Dominique Mulatier, Innovation Management Leader, Schneider Digital, Schneider Electric, France (2019)

> Too many challenges kill the challenge.

"At Schneider, a large international company, we constantly strive to innovate and introduce new products into the marketplace. For that purpose, we have developed a unique framework (business model), for which I am an example as Innovation Management Leader. We focus on what we call 'lean innovation'. This essentially means attempting to reduce 'time-to-profit' by engaging in three steps: (1) exploration, which includes brain storming and other such creative endeavors; (2) incubation, by which we test our ideas and prototypes both from an engineering point of view and with potential customers; and (3) scale-up, by which we finalize our design, set the production steps and, finally, market the product."

"This requires a substantial amount of training—knowledge is at the forefront of our leadership in the international marketplace. We train, coach, and assist in a variety of ways, including by furthering team building, value creation and, of course, innovation. Personally, I hold about 15 training sessions a year, complete with webinars and coaching, which include approximately ten people per group. Our three areas of focus are the technical aspects (we are, fundamentally, an engineering firm), marketing, and finance. Nowadays, there is just no way around the fact that project managers (often engineers themselves) must ally with marketers and check in with the finance department to ensure innovative ideas progress well and quickly without, however, sacrificing quality."

"We build consensus by methodically structuring our efforts, providing methodological support, grounding ourselves in shared experiences (both past and present), and focusing on a common objective. Unfortunately, we've noticed the younger generation has a hard time concentrating for long periods, and sometimes lacks tenacity in their efforts. Blame social media, I suppose!"

"When we seek innovation, we often emphasize new functionalities to resolve current or emerging problems our clients face. Over the years we have launched a large number of projects, and I think it is fair to say the sources of failure are generally the following:

1. An innovation that does not hold its promise(s),
2. A market that is not yet primed to accept the innovation,
3. A clientele that has been wrongly assessed, or
4. Margins that turn out to be too low despite initial forecasts."

"Of course, when we develop new products, we hypothesize market trends and desires. We also test them, but there is always a risk that either we have made a mistake or the market changes in a way that just could not be predicted. These hazards are often linked to human factors (people change), which add stakes to the initial challenge new product development represents."

"I'd say that eliminating challenge is essentially eliminating innovation."

6.2.4 Iron Triangle

6.2.4.1 D. Bourgoin, Project Manager, Canada (2016)

> As binding as it may seem to the project manager (each project being different from the previous one), standards give a base to build upon in terms of both quality assurance and sharpened project monitoring and control.

"When dealing with improving efficiency of project management practices within an organization, it is important to put some standards in place." It is difficult to evaluate the efficiency of a process if the project manager does not have a reference to compare… Standards help establish ways to execute the project that the management processes. Preliminary schedules and budgets, as well as a project charter presented in a standardized way, do simplify the analysis and the decision process of the project manager.

"Moreover, it may be time-consuming to establish a new set of documents to run a project (project charter, change requests, issue register, etc.). Having some pre-defined standards avoids 'reinventing the wheel'. Sharing information in a multi-site organization is often a considerable challenge; standards serve as facilitators for communication, monitoring, and reporting."

6.2.5 Key Consensus Factors (KCFs)

6.2.5.1 David Saulnier at A.F. Thériault, Boat Building Company, Nova Scotia, Canada (2015)[1]

> Each product is scrupulously crafted, much like a piece of art.

"My staff trusts me because I am easy to connect with; I am not saying easy to deal with—there is a difference. I am the boss, but my approach is convivial. I never hesitate to congratulate my staff for their dedication and hard work. We take a lot of pride in the Hammerhead series. This is something we have developed from scratch."

6.2.5.2 Thierry Bièvre, PDG, Élithis, France (2019)

> Sharing is the guiding force that allows us to overcome obstacles, and to realize our projects to everyone's satisfaction.

---

[1] For all testimonies ranging from 2015 to 2016 inclusively, see Mesly (2017).

"Our experience has taught us that a transdisciplinary approach is increasingly becoming the one asset project managers need in order to succeed. What favors this approach, you ask?"

"First, all stakeholders have to give a reason for their own actions, a common interest that binds everyone. Second, this has to be intimately related with the creation of value, particularly for our clientele and society as a whole. (Many of our architectural projects are eco-friendly.) Finally, we attract talent, and when talented people get together, a great group dynamic occurs; the result is always more than the sum of its parts."

"We see ourselves as talking the language of the future. Our engineers are explorers; they seek innovation in whatever they do. Furthermore, our quest for innovation never stands still or in silos."

6.2.6 Key Failure Factors (FFFs)

6.2.6.1 Éric Déry, Banque de développement, Banker, Gatineau, Canada (2015)

> We look for camouflaged details or motivations by way of the documentation, interviews, and visits to the facilities.

"Years of experience have taught me to detect possibilities of failure, which I can often judge by testing the trust level that exists between the key stakeholders, including us, the funders. We check the curriculum of the applicant, [their] credit and criminal (if any) background, past bankruptcies, and verify whether the project is overly optimistic."

6.2.7 An International Fair Dedicated to New Product Introduction in a New Market (2011–2019)

Students love this exercise of an international fair, and have for many years. We strongly recommend it to professors and tutors, as well as coaches, as it can be adapted to a work environment to encourage bonding between team members during training sessions. In the example below, students participated in a unique project whereby they prepared a booth for an international fair. (This we actually held at the school where the course in marketing feasibility of project is.) They developed a unique product, their own promotional campaign, and all the required material to ensure the targeted clients attending the fair could sign a contract for immediate (or future) purchases. The entire community was invited. Some professors and community members act as troublemakers, duty officers, eager clients, and so forth. The fair lasts for approximately three hours and requires roughly twelve weeks of intense work. After, the students write a full report on what they have done, learned, and what the other booths at the fair had to offer.

We provide some examples below of this great exercise built on experiential learning. We hereby report the experience as presented by the various groups of students[2] to give a feel for the actual experience, which emulates quite closely how things happen 'in real life,' that is, in the industry.

6.2.7.1 Diana Bracci, Carole Clain, Caterina Marino, and Celeste Schillani (2018)

We learned how to progressively proceed from theory to practice.

"During the first week, Week 1, we decided on the development of a simple and straightforward product. We were all interested in the food and beverage industry. Our first idea was to sell a small moka pot (a traditional Italian coffee machine) together with a specific blend of coffee. But after much discussion we decided to find a more innovative idea since the product already existed in the target market. So, we proceeded by conducting market research on the habits of coffee consumption of Canadian consumers."

"During the second week, Week 2, our idea got better defined: a premade, coffee-based cocktail. We started by creating the poster for the fair, then came up with the idea for the name: XXXX. (It stands for coffee and alcohol.) Once we chose the name we designed a logo and started to write a description of our product, listing the attributes (three different alcoholic coffee drinks: black, decaf, and creamy), the recipes, and the bottle sizes. We also prepared labels containing the nutritional value and alcoholic content of our product. We then continued by creating the marketing plan, and we made sure to have the necessary visas to travel abroad. Lastly, we checked the requirements to receive a Fairtrade and a vegan certification, and downloaded the application forms."

"During Week 3, we again changed the idea of the product in order to make it more interesting and innovative in the eyes of potential customers. The product changed from being a coffee-based cocktail to a kit for preparing coffee cocktails at home. The new product would be custom-made: The customer would choose, from our website, the type of spirit, the strength of the coffee, and whether or not to add milk. We would offer subscription packages that would be sent directly to the customers' doorstep, and would include all the pre-dosed ingredients. In the case of a first subscription/order we would also include a shaker and two glasses. We then made a list of primary and secondary attributes, choosing three main attributes: (1) custom-made, (2) easy, and (3) express." From this, we created a message made to attract the customers' attention: "XXXX, a coffee cocktail subscription

---

[2] We warmly thank the students for allowing us to use their work, and for participating in this great adventure.

kit: custom-made, easy quality, right at your home." This transformed into our promise: "Your custom-made bartender coffee experience, combining the best coffee and spirits, right at your door." We then changed a detail in our logo (from "the new coffee experience" to "the new cocktail experience") and started creating our selling sheet.

"During Week 4, we searched for the HS Code, yet found some difficulties in defining which was the right one because of various differentiations in the kinds of spirits. Second, we looked for some aids we could receive from the Canadian government, and found out we could seek assistance from various agencies (e.g., the Regional Aid, the France Label, the Prospecting Insurance). We wrote down a list of the documents we needed: Certificate of Origin, Carnet ATA for Canada (temporary import permit), photocopies of our passports, Electronic Travel Authorization, Pro forma invoice, packing list, Phytosanitary certificate, Form B3 (custom coding form), and an eTA Application. We downloaded them filling them up. We made further inquiries on the Canadian market, finding out not only their preferences on coffee and alcohol, but also information about the per-year expenditure on alcohol. Lastly, we continued with the creation of the selling sheet."

"During Week 5, we concentrated on the finalization of our logo, fair poster, and flyers. We completed the selling sheets, one for the general public and one for targeted customers, together with the sales forecast for the fair and for the next three years. We then worked on the cost sheet and the coffee cocktail trial tests. We drafted our segmentation and our positioning and realized we were very differentiated from our competitors. For positioning, we decided to include different gifts (cigars, chocolates, and sweets) according to different segments. We started buying some things for a booth: a shaker kit (to show as an example what we provide in the box) and a little vase for smelling coffee."

"During Week 6 we prepared an Agency Agreement, as well as the letter of intent for the fair. We finalized our segmentation with more-defined categories. We edited our P&L account according to the product lifecycle, adding more realistic expenses in the first year in order to make it negative. Lastly, we prepared business cards and invitations for the French authorities in Canada."

"As of Week 7 we had all our documents ready, so we focused on the booth. We bought a silver tablecloth, some lights, and started thinking about the position of the different elements on the table (computers, price list, gifts, coffee containers). We also looked for some academic articles that could support our products. Since we promote responsible alcohol consumption, we found one supporting our point of view. In addition, we translated the PPT presentation since we understood a large bulk of the fair's public would be French-speaking."

"Week 8 was the fair, and turned out to be a real success! To conclude, the international fair organization allowed us to have a concrete experience and to understand what will be expected of us in the foreseeable future. Through this experience we learned not only about the main concepts of project management, but also how to work cohesively within an international team. Since the course was structured by stages, we learned how to progressively proceed from theory to practice, from preparing documents to designing a booth, and gained some fundamental skills on the subject. This course made us understand how important it is to put into practice the theory we learn, making the transition from university to professional life smoother. Overall, it was both a fun and educational experience that allowed us to discover and exploit our potential."

6.2.7.2 Martina Capirossi, Marta Forlano, Maria Iliescu, and Lucas Slegers Tiago (2018)

Working with people with different backgrounds and cultures boosted the ideation and challenging process.

"All members of the group had previous marketing knowledge. The international fair gave us the possibility to transform this theoretical knowledge into practice within an international context. First, we studied the factors that influence the buying behavior of consumers in the Canadian market and, by doing so, we started understanding the process of internationalization. Second, we identified trends in order to understand differences across cultures. Keeping that in mind during the development of our product, we focused on the concept of brand value and how this is proposed. We then constituted it within an international context. Taking into consideration and analyzing our competitors we proposed a product with added value, namely, one with an educational purpose. We leveraged the characteristics to gain a competitive advantage over existing board game developers; indeed, the increasing competition requires managers to pay attention to brand meanings worldwide."

"Our team consisted of four people aged between 21 and 26, from different nationalities. This project gave us the possibility to work in an international environment without moving out of France! Working with people from different backgrounds and cultures boosted the ideation and creative process. The differences across countries became clear during the working process. The team members managed to distribute and organize tasks to merge and fully exploit each other's competencies. By doing so, we all compensated for our individual weaknesses. Finally, the group gained a logical understanding of differences across the European and Canadian markets."

"In the pre-fair booth preparation, on the day of the fair, our time management skills were essential due to the lack of time at disposal. Indeed, we managed to set up the booth in time and had everything settled down properly. During the setup, however, a problem emerged with our neighbors. This was due to the fact the space between booths was limited, and organizational issues arose. In the end, we found a compromise. This allowed us to experience what taking part in an actual fair means."

"Regarding the roles assigned to each person, three members of the team alternated between acting as customers and dealers selling our pitch. The other member was in charge of collecting business contacts and drew people toward the booth. The face-to-face dealings with potential customers gave us the possibility to practice, and improve, communication and community management."

"Overall, the project strengthened our team-building and analytical skills as well as our time management, which was fundamental for the whole project. We practiced crafting and product innovation skills because we invented the whole game from scratch: from the box to its contents. We created a tangible sample of the game, and we had to assemble the fair booth."

6.2.7.3 *Valentina Cijan, Micaela Escribano Bernard, Kristina Eseola, and Daniela Riccio (2018)*

Every project arises from a single idea.

"A single idea can be triggered by a proposal or a request, a need or a desire. Our trigger was the international fair class project; our motivation the desire to create a new product that fit into the Canadian market. Putting together our prior knowledge and traditions, we came up with several ideas related to food and beverage. These ideas were evaluated, designed, redesigned, and tested according to several analyses of the Canadian market and society. Our team worked really hard on the characteristics and attributes of the product to finally present and deliver the most innovative and high-quality commodity into a complex and exigent market. All aspects of our proposal were carefully taken into consideration, ranging from the origins of the raw materials and the production stages up to the shipping and delivery methods."

"There are things that can go wrong during an international fair. To be responsive to these, our company designed a contingency plan for: (1) problems at the border, such as visas; (2) no electricity in the fair compound; (3) flight delays; (4) bad weather; (5) forgetting important documentation that may be requested on site; (6) customs issues in regard to products exported/imported; (7) computers not working; (8) our product being out of stock; (9) bad positioning of the booth; (10) low attendance; (11) lack of interest from the attendants; (12) incorrectly building the booth; (13) problems with nearby booths; (14) allergies or illnesses of the booth presentors; (15) spacing (is there enough to deliver our speech or presentation); (16) short on staff; (17) sample delays; (18) jet lag; (19) traffic jams and late arrival to the event; (20) lost luggage; (21) family problems; and (22) language problems."

CHAPTER 6, CLASS EXERCISE #1:

Among all the testimonials, which one speaks to you most? Why?

6.3 Conclusion

The testimonials we provided placed the theory discussed throughout this book in a pragmatic perspective. We collected testimonials issued from a wide variety of economic sectors, projects, and cultures. We believe they provide readers an opportunity to "live" the theory, or else to recognize that others have lived what we have experienced. They also connect marketing and project management in a pragmatic light, one that theory cannot fully express. Together with the various excerpts we provided in Chapter 5 from different projects the world over, and the exercises that we give during our seminars, these testimonials enrich the field where marketing and project management meet.

Through these various forms of expression, both academics and practitioners will feel the value of the concepts presented in this book. We paid particular attention to discussing concepts marketers and project managers use in their everyday, working lives. We encourage readers to discuss their own marketing and project management experiences, and to exchange ideas.

Conclusion

An artistic draft is in the foreground of the image of an actual building in the background.

We have covered a fair bit of ground in this book. We examined marketing and the 6Ps of marketing management. We looked at project management and the 6Ps of strategic management. We joined the 4Ps of marketing to the 4Ps of feasibility of projects, and showed how the iron triangle of timeline, budget, and norms of quality concerns both marketing and project managers. We defined projects and expanded on prefeasibility and feasibility analyses, and developed our thoughts with respect to marketing feasibility of projects. We showed an increasing number of marketing and project managers connect with potential clients/end users in the development and realization of projects. We outlined a substantial number of argu-able errors we believe need to be discussed in the context of where marketing and project management meet. We proposed exercises that actually assist marketing and project managers in their daily tasks. This is why we paid attention to challenging readers with, for example, listing attributes of brands they recognize instantly and that occupy their minds, hearts, and wallets. Readers can simply transpose this effort to projects, and they will notice that the same constructs and managerial concepts take place.

We discussed some of the challenges both marketing and project managers faced in the past, as well as now. We ventured into the future and predicted that marketing and project management will continue evolving and, more particularly, co-evolving. This will take place in a number of ways.

Corporate social responsibility (CSR) is fast becoming an inescapable theme of marketing and project management. We are still a long way before corporations "behave" and limit their delinquent behaviors, or hide them behind official good governance principles that in the end cover harmful agendas. Many institutions and corporations can actually afford to rebuff their own ethical behaviors, because it is often too costly and too time-consuming to force them to act in good faith. Yet the mere fact that marketers, project managers, and clients cooperate will serve to provide better safeguards against deviant behaviors. Why would future end users approve of behaviors, marketing techniques, or dubious project management practices that would, in the end, deflate their dreams and somehow damage their lives in one way or another?

Both managers will have to become more flexible in their approach to management and to working together. We find the Project Management Book of Knowledge (PMBOK) to be a very stiff manual, one that has great value but that nevertheless would gain in understanding better the "science" of feasibility of projects, the role of work psychodynamics, and the increasing necessity to be flexible or, put in typical project management terms, **agile**. Academics define agile projects in a number of ways. In reality, agility simply means we can no longer set strict guidelines that cannot deviate from a preset, engineering-like type path of development. Instead, one must acknowledge that every project faces unexpected risks and points of vulnerability (POVs) and requires, therefore, flexibility.

Another development that will certainly occupy more of the marketing and project management space is the use of **virtual reality** and artificial intelligence. Clients can "live" the project and its deliverable by way of wearing special 3D glasses, complemented with sounds and even odors. We, as authors and researchers, have our own virtual reality software that we use for testing behaviors in different decision-making circumstances. Virtual reality is a fun and exciting tool and can save a substantial amount of money, especially in assessing the match between the anticipated deliverable and the clients' needs.

Lean innovation is part of the underlying principles that guide the present book. Through better assessment of the project, and more articulated relationships between stakeholders (including marketers, project managers, and clients), better prototyping leads to faster time to market.

In terms of managerial considerations, there are a certain number of general principles that guide project management and, certainly, the marketing of projects. A project is not feasible if the positive forces (which play in favor of the project; i.e., strengths and opportunity) are fewer than the negative forces (risks and vulnerabilities). The higher the number of dependencies between the project's tasks, the more vulnerable the project is; the causal chain becomes more precarious as intricacies between its constituents increase. Also, the higher the number of POVs and the weaker the remedial actions are, the less the project is feasible. Obviously, the more wrangling among stakeholders that is intense, frequent, and cover critical issues, the less feasible the project is. The more complex a project is, the more likely it contains POVs; hence, the more the probability of failure exists.

An inescapable concern to managers, researchers, and scientists keen on innovation is **techno-predation**, which is "the sudden and unwelcome appropriation or use of a new technology by a stakeholder in an innovation network to the detriment of the creator of said technology."[1] There is nothing new with techno-predation. Since the beginning of World War II (for example), market winners are those who have taken a technological lead—in many cases, with the intervention of technological borrowing or theft. Academics are not protected more than industrial researchers and this endemic problem affects, and will continue to affect, particular types of projects such as IT, pharmaceutical, and military. Regularly, scientists working at universities see their innovations or articles stolen by colleagues or competitors. Many universities are ill-equipped to deal with this phenomenon, even though they claim they provide a safe haven for research. The situation is not any better within private organizations, where plagiarism and intellectual rights infringement are an everyday problem. Innovation is plagued with POVs, and as time goes on, experts become more astute in getting around security measures; this problem will only grow.

We believe innovation, which is one of the defining elements of projects and a key component of marketing strategy, needs a set of values that marketers and project managers could adopt for the benefit of all projects, scientific or artistic alike. Perhaps naively, we posit these **human values** for governing innovation are:

1. Candor, or the attitude by which we remain humble in the face of everything there is still to discover. This can be read as fairness from the psychodynamic model's point of view.
2. Friendship, or the capacity to create constructive bonds given that innovation becomes increasingly complex and transdisciplinary. (Read, cooperation.)
3. Integrity, or the adherence to strict and ethically sound guiding principles that serve the interests of all, in particular to subdue techno-predation. (Read, trust.)

---

[1] Mesly (2018).

4. Openness, or the willingness to identify hidden needs and new opportunities. (Read, transparency and control.)
5. Passion, or the relentless motivation to bring value to others and to self.
6. Recognition, or the act of acknowledging everyone's input. (Read, reward.)

Let there be said that marketing and project managers should continue to develop a mutually beneficial understanding, preferably by seeking the resourceful inputs of end users. Let there be said that such co-evolution fosters human values, those that make our societies a better and safer place not only for us but for generations to come. By digging deep into our endeavors and analyzing whether our projects are truly feasible, we accept and recognize our vulnerabilities and are able to address them before they emerge and become unmanageable—before they threaten the project's realization. This book is by no means the end of such efforts; it is merely a modest attempt to productively merge two disciplines: marketing and project management. Nowadays, the two certainly have no choice but to cohabit in concerns to any project worldwide. Companies will continue to seek, at an increasing rate, marketing experts who understand and manage projects, and project managers who can enrich their own actions by way of better understanding and applying the principles of marketing relevant to their endeavors.

We trust this book, if anything, has ignited the desire for each group of managers to work together, and to know more about other fields of expertise.

Mind Teasers

Readers may use the mind teasers as questions in preparation for an examination or quiz.

1. In the context of project management, define
   a. agility,
   b. lean innovation,
   c. techno-predation, and
   d. the tool of virtual reality.
2. List, and briefly discuss,
   a. human values that are assumed to promote innovation and
   b. some managerial considerations discussed throughout this text.

Bibliography

Aaltonen, K., Kujala, J., Havela, L., and Savage, G. (2015). Stakeholder dynamics during the project front-end: the case of nuclear waste repository projects. *Project Management Journal* 46(6), 15–41.

Abednego, M., and Ogunlana, S.O. (2006). Good project governance for proper risk allocation in public–private partnerships in Indonesia. *International Journal of Project Management* 24, 622–634.

Ahola, T., Laitinen, E., Kujala, J., and Wikström, K. (2008). Purchasing strategies and value creation in industrial turnkey projects. *International Journal of Project Management* 26(1), 87–94.

Akintoye, A. (2000). Analysis of factors influencing project cost estimating practice. *Construction Management and Economics* 18(1), 77–89.

Akintoye, A., and MacLeod, M. (1997). Risk analysis and management in construction. *International Journal of Project Management* 15(1), 31–38.

Alderman, N., and Ivory, C. (2007). Partnering in major contracts: Paradox and metaphor. *International Journal of Project Management* 25(1), 386–393.

Alvarez, J.F.A., Pustina, A., and Hällgren, M. (2011). Escalating commitment in the death zone: New insights from the 1996 Mount Everest disaster. *International Journal of Project Management* 29, 971–985.

American Psychiatric Association. (2013). *Diagnosis and statistical manual of mental disorders (DSM)*, 5th edition. Washington, DC: American Psychiatric Association.

Andersen, E., Söderlund, J., and Vaagaasar, A. (2010). Projects and politics: Exploring the duality between action and politics in complex projects. *International Journal of Management and Decision Making* 11(2), 121–139.

Aritua, B., Smith, N.J., and Bower, D. (2009). Construction client multi-projects: A complex adaptive systems perspective. *International Journal of Project Management* 27, 72–79.

Baccarini, D. (1996). The concept of project complexity—A review. *International Journal of Project Management* 14(4), 201–204.

Bales, R.F. (1950). *Interaction process analysis: A method for the study of small groups*. Reading Mass: Addison-Wesley.

Bass, B. (1960). *Leadership, psychology and organizational behavior*. New York: Holt, Rinehart and Winston.

Belassi, W., Kondra, A.Z., and Tukel, O.I. (2007). The effects of organizational culture. *Project Management Journal* 38(4), 12–24.

Berssaneti, F.T., and Carvalho, M.M. (2015). Identification of variables that impact project success in Brazilian companies. *International Journal of Project Management* 33, 638–649.

Besner, C., and Hobbs, B. (2013). Contextualized project management practice: A cluster analysis of practices and best practices. *Project Management Journal* 44(1), 17–34.

Bhattacharya, P., and Mehta, K.K. (2000). Socialization in network marketing organizations: Is it cult behavior? *Journal of Behavioral and Experimental Economics*, 29(4), 361–374.

Boston Consulting Group: www.bcg.com/

Bowlby, J. (1973). *Attachment and loss. Separation: Anxiety and anger*, Vol. 2. New York: Basic Books.

Brady, T., and Davies, A. (2010). From hero to hubris: Reconsidering the project management of Heathrow's Terminal 5. *International Journal of Project Management* 28, 151–157.

Brockhoff, K. (2003). Customers' perspectives of involvement in new product development. *International Journal of Technology Management* 26(5/6), 464–481.

Brockner, J. (1992). The escalation of commitment to a failing course of action: Toward theoretical progress. *Academy of Management Review* 17(1), 39–61.

Brockner, J., Nathanson, S., Friend, A., Harbeck, J., Samuelson, C., Houser, R., Bazerman, M.H., and Rubin, J.Z. (1984). The role of modeling processes in the "knee deep in the big muddy" phenomenon. *Organizational Behavior and Human Performance* 33(1), 77–99.

Bruelious, N., Flyvbjerg, B., and Rothengatter, W. (1998). Big decisions, big risks: Improving accountability in mega projects. *International Review of Administrative Sciences* 64(3), 423–440.

Bryde, D.J., and Robinson, L. (2005). Client versus contractor perspectives on project success criteria. *International Journal of Project Management* 23, 622–629.

Buchanan, D.A. (1991). Beyond content and control: Project vulnerability and the process agenda. *International Journal of Project Management* 9(4), 233–239.

Butler, R.S., DeBower, H.F., and Jones, J.G. (1914). *Marketing methods and salesmanship*. New York: Alexander Hamilton Institute New York.

Carron, A.V., Widmeyer, W.N., and Brawley, L.R. (1988). Group cohesion and individual adherence to physical activity. *Journal of Sport and Exercise Psychology* 10, 127–138.

Chang, C.Y., and Ive, G. (2011). Selecting procurement systems for capital projects: A transaction cost perspective. In: *Advances in business and management*, Vol. 2, ed. W.D. Nelson, pp. 125–139. New York: Nova Science Publishers, Inc.

Chang, W., and Taylor, S.A. (2016). The effectiveness of customer participation in new product development: A meta-analysis. *Journal of Marketing* 80, 47–64.

Chen, C.Y. (2011). Managing projects from a client perspective: The concept of the meetings-flow approach. *International Journal of Project Management* 29(6), 671–686.

Cherns, A.B., and Bryant, D.T. (1984). Studying the client's role in construction management. *Construction Management and Economics* 2(2), 177–184.

Chih, Y.Y., and Zwikael, O. (2015). Project benefit management: A conceptual framework of target benefit formulation. *International Journal of Project Management* 33(1), 352–362.

Clark, K.B., and Wheelwright, S.C. (1992). Organizing and leading 'heavyweight' development teams. *California Management Review* 34(3), 9–28.

Consoli, G.G.S. (2006). Conflict and managing consortia in private prison projects in Australia" private prison operator responses. *International Journal of Project Management* 24(1), 75–82.

Cooper, R.G. (1979). The dimensions of industrial new product success and failure. *Journal of Marketing* 43, 93–103.

Cova, B., and Cova, V. (2002). Tribal marketing: The tribalisation of society and its impact on the conduct of marketing. *European Journal of Marketing* 36(5–6), 595–620.

Cova, B., Ghauri, P., and Salle, R. (2002). *Project marketing: Beyond competitive bidding.* Chichester: John Wiley & Sons.

Coviello, N.E., and Joseph, R.M. (2012). Creating major innovations with customers: Insights from small and young technology firms. *Journal of Marketing* 76, 87–104.

Creasy, T., and Anantatmula, V.S. (2013). From every direction: How personality traits and dimensions of project managers can conceptually affect project success. *Project Management Journal* 44(6), 36–51.

Crespin-Mazet, F., and Ghauri, P. (2007). Co-development as a marketing strategy in the construction industry. *Industrial Marketing Management* 36, 158–172.

Dalal, A.F. (2012). *The 12 pillars of project excellence: A lean approach to improving project excellence.* Florida: Taylor & Francis Group.

Davies, A., and Brady, T. (2000). Organizational capabilities and learning in complex product systems: Towards repeatable solutions. *Research Policy* 29(7/8), 931–953.

de Gaulle, G. (1954). *Mémoires de guerre: L'Appel 1940–1942.* Paris: Le Livre de Poche, Librairie Plon.

Dingle, J. (1985). Project feasibility and manageability. *Project Management* 3(2), 94–103.

Doloi, H.K. (2011). Understanding stakeholders' perspective of cost estimation in project management. *International Journal of Project Management* 29(5), 622–636.

Dulewicz, V., and Herbert, P. (1999). Predicting advancement to senior management from competencies and personality data: A seven-year follow-up study. *British Journal of Management* 10(1), 13–22.

Dvir, D., Sadeh, A., and Pines, A.M. (2006). Projects and project managers: The relationship between project managers' personality, project types, and project success. *Project Management Journal* 37(5), 36–48.

Ekman, P. (1999). Basic emotions. In: *Handbook of cognition and emotion*, ed. T. Dalgleish and M. Power. New York: John Wiley and Sons, Ltd.

Elton, E.J., Gruber, M.J., Brown, S. J., and Goetzmann, W.N. (2011). *Modern portfolio theory and investment analysis*, 8th edition. New York: John Wiley & Sons Ltd.

Ernst, H., Hoyer, W.D., and Rübsaamen, C. (2010). Sales, marketing, and research-and-development cooperation across new product development stages: Implications for success. *Journal of Marketing* 74(5), 80–92.

Flyvbjerg, B. (2005). Policy and planning for large infrastructure projects: Problems, causes, cures. *Environment and Planning B: Planning and Design* 34, 578–597.

Forbes, www.forbes.com/.

Gaddis, P.O. (1959). The project manager. *Harvard Business Review* 37(3), 89–97.

Goldberg, L.R. (1990). An alternative "description of personality": The big-five factor structure. *Journal of Personality and Social Psychology* 59(6), 1216–1228.

Graham, J.H. (1996). Machiavellian project managers: Do they perform better? *International Journal of Project Management* 14(2), 67–74.

Griffin, A., and Hauser, J.R. (1992). Patterns of communications among marketing, engineering and manufacturing: A comparison between two new product teams. *Management Science* 38(3), 360–374.

Grönroos, C. (2011). A service perspective on business relationships: The value creation, interaction and marketing interface. *Industrial Marketing Management* 40, 240–247.

Gruner, K.E., and Homburg, C. (2000). Does customer interaction enhance new product success? *Journal of Business Research* 49(1), 1–14.

Guimond, S., and Dambrun, M. (2002). When prosperity breeds intergroup hostility: The effects of relative gratification on prejudice. *Personality and Social Psychology Bulletin* 28, 900–912.

Gupta, S., and Zeithaml, V. (2006). Customer metrics and their impact on financial performance. *Marketing Science* 25(6), 718–739.

Gutierrez, O., and Friedman, D.H. (2005). Managing project expectations in human services information systems implementations: The case of homeless management information systems. *International Journal of Project Management* 23, 513–523.

Hadjikhani, A. (1996). Project marketing and the management of discontinuity. *International Business Review* 5(3), 319–336.

Hammer, W.C., and Yukl, G.A. (1977). The effectiveness of different offer strategies in bargaining. In: *Negotiations: Social-psychological perspectives*, ed. D. Druckman. London: Sage Publications.

Hansen, C.H., and Hansen, R.D. (1988). Finding the face in the crowd: An anger superiority effect. *Journal of Personality and Social Psychology* 54, 917–924.

Haumann, T., Güntürkün, P., Schons, L.M. and Wieseke, J. (2015). Engaging customers in coproduction processes: How value-enhancing and intensity-reducing communication strategies mitigate the negative effects of coproduction intensity. *Journal of Marketing* 79, 17–33.

Haverila, M.J., and Fehr, K. (2016). The impact of product superiority on customer satisfaction in project management. *International Journal of Project Management* 34, 570–583.

Heravi, A., Coffey, V., and Trigunarsyah, B. (2015). Evaluating the level of stakeholder involvement during the project planning processes of building projects. *International Journal of Project Management* 33, 985–997.

Herzberg, F., Mausner, B., and Snyderman, B.B. (1959). *The motivation to work*. New York: John Wiley.

Hillman, A.J., and Keim, G.D. (2001). Shareholder value, stakeholder management, and social issues: What's the bottom line? *Strategic Management Journal* 22(2), 125–139.

Hirsley, M. (April 28, 1985). To Southerners, New Coke Just Isn't It. Chicago Tribune.

Hobday, M. (2000). The project-based organization: An ideal form for managing complex products and systems? *Research Policy* 29(7/8), 871–894.

Homburg, C., Schwemmle, M., and Kuehnl, C. (2015). New product design: Concept measurement and consequences. *Journal of Marketing* 79, 41–56.

http://environprojects.en.hisupplier.com/.

http://mria-arim.ca/fr/a-propos-de-larim/normes/code-de-deontologie-des-membres,

https://fr.scribd.com/doc/5924351/The-Nstp-Project-Cesar-Community-Extension-Services-through-Action-and-Research.

https://lam.can-am.brp.com/on-road/owners/spyder-blog/the-inside-scoop-on-the-7th-annual-owners-event.html.

Hughes, M. (2013). The Victorian London sanitation projects and the sanitation of projects. *International Journal of Project Management* 31, 682–691.

Hugo, V. (2001). Quatrevingt-treize. Édition de Bernard Leuilliot, Le Livre de poche (classique).

Hwang, B.G., Zhao, X., and Goh, K.J. (2014). Investigating the client-related rework in building projects: The case of Singapore. *International Journal of Project Management* 32, 698–708.

Idrissou, L., van Paassen, A., Aarts, N., Vodouhè, S., and Leeuwis, C. (2013). Trust and hidden conflict in participatory natural resources management: The case of the Pendjari National Park (PNP) in Benin. *Forest Policy and Economics* 27, 65–74.

International Monetary Fund (IMF) (2009). www.imf.org.

Jacques Cartier and Champlain Bridges Incorporated, JCCBI. (2011). Prefeasibility study concerning the replacement of the existing Champlain Bridge.

Jónsson, H.R. (2012). Feasibility analysis procedures for public projects in Iceland. *Thesis for the degree of master of science in construction management. Reykjavík University, Iceland.*

Jørgensen, M., and Sjøberg, D.I.K. (2004). The impact of customer expectation on software development effort estimates. *International Journal of Project Management* 22, 317–325.

Kadefors, A. (1995). Institutions in building projects: Implications for flexibility and change. *Scandinavian Journal of Management* 11(4), 395–408.

Kähr, A., Nyffenegger, B., Krohmer, H., and Hoyer, W.D. (2016). When consumers wreak havoc on your brand: The phenomenon of consumer brand sabotage. *Journal of Marketing* 80, 25–41.

Kaulio, M.A. (1998). Customer, consumer and user involvement in product development: A framework and a review of selected methods. *Total Quality Management* 9(1), 141–149.

Kilmann, R.H., and Thomas, K.W. (1975). Interpersonal conflict-handling behavior as reflections of Jungian personality dimensions. *Psychological Reports* 37(3), 971–980.

Kloppenborg, T., and Opfer, W. (2002). The current state of project management research: Trends, interpretations, and predictions. *Project Management Journal* 33(2), 5–18.

Landwehr, J.R., Wentzel, D., and Herrmann, A. (2013). Product design for the long run: Consumer responses to typical and atypical designs at different stages of exposure. *Journal of Marketing* 77, 92–107.

Lecoeuvre-Soudain, L., Deshayes, P., and Tikkanen, H. (2009). Positioning of the stakeholders in the interaction project management: Project marketing: A case of a co-constructed industrial project. *Project Management Journal* 40(3), 34–46.

Lu, S., and Yan, H. (2007). An empirical study on incentives of strategic partnering in China: Views from construction companies. *International Journal of Project Management* 25, 241–249.

Mackenzie, W. (2011). *Large capital projects benchmarking.* Scotland (Edinburgh): Wood Mackenzie Research.

Magdol, L., Moffitt, T.E., Caspi, A., and Silva, P.A. (1998). Developmental antecedents of partner abuse: A prospective–longitudinal study. *Journal of Abnormal Psychology* 107(3), 375–389.

Mäkilouko, M. (2004). Coping with multicultural projects: The leadership styles of Finnish project managers. *International Journal of Project Management* 22, 387–396.

Maloney, W.F. (2002). Construction product/service and customer satisfaction. *Journal of Construction Engineering Management* 128(6), 522–529.

Mantel, S.J., Meredith, J.R., Shafer, S.M., and Sutton M.M. (2011). *Project management in practice.* Hoboken, NJ: John Wiley & Sons, Inc.

McClelland, D. (1988). *Human motivation.* Cambridge: Cambridge University Press.

Meredith, J.R., and Mantel, S.J. (2009). *Project management: A managerial approach*, 7th edition. New Jersey: John Wiley & Sons Inc.

Mesly, O. (2010). Voyage au cœur de la prédation entre vendeurs et acheteurs: Une nouvelle théorie en vente et marketing. *Doctoral thesis. Université de Sherbrooke. Bibliothèque nationale,* 505 pages.

Mesly, O. (2011). *Une façon différente de faire de la recherche en vente et marketing. Presses de l'Université du Québec*. Québec: Presses de l'Université du Québec, 202 pages.

Mesly, O. (2015a). *Creating models in psychological research*. Cham: Springer International Publishing.

Mesly, O. (2015b). The role of physical distance in six interpersonal core competencies in international construction projects. *International Journal of Project Management* 33, 1425–1437.

Mesly, O. (2017). *Project feasibility: Tools for uncovering points of vulnerability*. New York, NY: Taylor and Francis.

Mesly, O. (2018). Techno-predation: A contemporary phenomenon jeopardizing innovation networks all over the world. *Journal of Small Business and Entrepreneurship* 31(6), 483–493.

Meyer, W.G. (2014). The effect of optimism bias on the decision to terminate failing projects. *Project Management Journal* 45(4), 7–20.

Mikulincer, M., and Shaver, P.R. (2007). *Attachment in adulthood: Structure, dynamics, and change*. New York: The Guilford press.

Milgram, S. (1974). *Soumission à l'autorité*. Paris: Calmann-Lévy.

Moenaert, R.K., Souder, W.E., de Meyer, A., and Deschoolmeester, D. (1994). R&D: Marketing integration mechanisms, communication flows, and innovation success. *The Journal of Product Innovation Management* 11(1), 31–45.

Moreno, J.L. (1969). *Les fondements de la sociométrie*. Paris: PUF.

Morris, P. (2013). Reconstructing project management reprised: A knowledge perspective. *Project Management Journal* 44(5), 6–23.

Morris, P.W.G. and Pinto, J.K. (2004). *The Wiley guide to managing projects*. New Jersey: John Wiley & Sons Inc.

Morris, P.W.G. (1994). *The management of projects*. London: Thomas Telford Services Ltd.

Morris, P.W.G. (2011). A brief history of project management. In: *The Oxford handbook of project management*, eds. P.W.G. Morris, J.K. Pinto, and J. Soderlund, pp. 15–36. Oxford: Oxford University Press.

Muller, R., and Jugdev, K. (2012). Critical success factors in projects. *International Journal of Managing Projects in Business* 5(4), 757–775.

Barrick, R.M., Mount, K.M., and Judge, A.T. (2001). "Personality and Performance at the Beginning of the New Millennium: What Do We Know and Where Do We Go Next? Personality And Performance, 9(1), 9-30.

O'Shaugnessy, W. (1992). *La faisabilité de projet*. Trois-Rivières, Canada: Les Éditions SMG.

Ohlendorf, A. (2001). Conflict resolution in project management. Information Systems Analysis MSIS 488, www.umsl.edu/~sauterv/analysis/488_f01_papers/Ohlendorf.htm.

Ouchi, W. (1979). A conceptual framework for the design of organizational control mechanisms. *Management Science* 25(9), 833–848.

Oyedele, L.O. (2013). Analysis of architects' demotivating factors in design firms. *International Journal of Project Management* 31, 342–354.

Papadopoulos, T., Ojiako, U., Chipulu, M., and Lee, K. (2012). The criticality of risk factors in customer relationship management projects. *Project Management Journal* 43(1), 65–76.

Parkhe, A. (1993). "Messy" research, methodological predispositions, and theory development in international joint ventures. *The Academy of Management Review* 18(2), 227–268.

Patanakul, P., Kwak, Y.H., Zwikael, O., and Liu, M. (2016). What impacts the performance of large-scale. *International Journal of Project Management* 34, 452–466.

Payne, A., Storbacka, K., and Frow, P. (2008). Managing the co-creation of value. *Journal of the Academy of Marketing Science* 36(1), 83–96.

Peled, M., and Dvir, D. (2012). Towards a contingent approach of customer involvement in defence projects: An exploratory study. *International Journal of Project Management* 30, 317–328.

Pinto, J.K. (2014). Project management, governance, and the normalization of deviance. *International Journal of Project Management* 32(3), 376–387.

Pinto, J.K., and Prescott, J.E. (1988). Variations in critical success factors over the stages in the project life cycle. *Journal of Management* 14(1), 5–18.

PM Network Magazine. (June 15). *PM network: Professional magazine of the project management institute*, Vol. 29, nbr 7. Newton Square, PA: PMI, p. 17.

PMBOK 5. (2013). *A guide to the project management body of knowledge*, 5th edition. Newtown Square, PA: Project Management Institute.

PMBOK 6. (2017). *A guide to the project management body of knowledge*, 6th edition. Newtown Square, PA: Project Management Institute.

Polesie, P. (2013). The view of freedom and standardisation among managers in Swedish construction contractor projects. *International Journal of Project Management* 31(2), 299–306.

Ratcheva, V. (2009). Integrating diverse knowledge through boundary spanning processes: The case of multidisciplinary project teams. *International Journal of Project Management* 27, 206–215.

Ringuest, J., and Graves, S. (1999). Formulating R&D portfolios that account for risk. *Research Technology Management* 42(6), 40–43.

Ritter, T., and Gemünden, H.G. (2003). Network competence: Its impact on innovation success and its antecedents. *Journal of Business Research* 56, 745–755.

Rogers, E.M. (1962). *Diffusion of innovation*. Glencoe: Free Press of Glencoe.

Ross, J., and Staw, B.M. (1986). Expo 86: An escalation prototype. *Administrative Science Quarterly* 31, 274–297.

Rubin, J., and Brockner, J. (1975). Factors affecting entrapment in waiting situations: The Rosencrantz and Guildenstern effect. *Journal of Personality and Social Psychology* 31(6), 1054–1063.

Ruuska, I., and Teigland, R. (2009). Ensuring project success through collective competence and creative conflict in public–private partnerships: A case study of Bygga Villa, a Swedish triple helix e-government initiative. *International Journal of Project Management* 27, 323–334.

Ruuska, I., Artto, K., Aaltonen, K., and Lehtonen, P. (2009). Dimensions of distance in a project network: Exploring Olkiluoto 3 nuclear power plant project. *International Journal of Project Management* 27, 142–153.

Sanderson, J. (2012). Risk, uncertainty and governance in megaprojects: A critical discussion of alternative explanations. *International Journal of Project Management* 30(4), 432–443.

Sauser, B., Reilly, R.R., and Shenhar, A. (2009). Why projects fail? How contingency theory can provide new insights: A comparative analysis of NASA's mars climate orbiter loss. *International Journal of Project Management* 27(7), 665–679.

Schulze, C., Skiera, B., and Wiesel, T. (2012). Linking customer and financial metrics to shareholder value: The leverage effect in customer-based valuation. *Journal of Marketing* 76, 17–32.

Shi, Q., Liu, Y., Zuo, J., Pan, N., and Ma, G. (2015). On the management of social risks of hydraulic infrastructure projects in China: A case study. *International Journal of Project Management* 33, 483–496.

Shore, B. (2008). Systematic biases and culture in project failures. *Project Management Journal* 39(4), 5–16.

Smyth, H., and Lecoeuvre, L. (2015). Differences in decision-making criteria towards the return on marketing investment: A project business perspective. *International Journal of Project Management* 33, 29–40.

Snyder, M.L., and Uranowitz, S.W. (1978). Reconstructing the past: Some cognitive consequences of person perception. *Journal of Personality and Social Psychology* 36, 941–950.

Srivastava, K. R., Shervani, T. A., and Fahey, L. (1998). Market-based assets and shareholder value. *Journal of Marketing* 62, 2–18.

Stephens, J.P., and Carmeli, A. (2016). The positive effect of expressing negative emotions on knowledge creation capability and performance of project teams. *International Journal of Project Management* 34, 862–873.

Sutherland, V.J., and Davidson, M.J. (1989). Stress among construction site managers: A preliminary study. *Stress Medicine* 5, 221–35.

Sydow, J., Schreyögg, G., and Koch, J. (2009). Organizational path dependence: Opening the black box. *Academy of Management Review* 34(4), 689–709.

Tabassi, A.A., and Abu Bakar, A.H. (2009). Training, motivation, and performance: The case of human resource management in construction projects in Mashhad, Iran. *International Journal of Project Management* 27, 471–480.

Thomke, S., and von Hippel, E. (2002). Customers as innovators: A new way to create value. *Harvard Business Review* 80(4), 74–81.

Tikkanen, H., Kujala, J., and Artto, K. (2007). The marketing strategy of a project-based firm: The four portfolios framework. *Industrial Marketing Management* 36, 194–205.

Todorov, A., and Engell, A.D. (2008). The role of the amygdala in implicit evaluation of emotionally neutral faces. *Scan* 3, 303–312.

Toledo, C., Leija, L., Munoz, R., Vera, A., and Ramirez, A. (2009). Upper limb prostheses for amputations above elbow: A review. In: *2009 pan American health care exchanges*. Piscataway: IEEE, pp. 16–20.

Transparency International: www.transparency.org/

Troyat, H. (1951). *La tête sur les épaules*. Paris, Librairie Plon.

Tysseland, B.E. (2008) Life cycle cost based procurement decisions: A case study of Norwegian defence procurement projects. *International Journal of Project Management* 26, 366–375.

Van Goozen, S.H.M., Matthys, W., Cohen-Kettenis, P.T., Buitelaar, J.K., and Van Engeland, H. (2000). Hypothalamic-pituitary-adrenal axis and autonomic nervous system activity in disruptive children and matched controls. *Journal of the American Academy of Child and Adolescent Psychiatry* 39(11), 1438–1445.

van Marrewijk, A. (2007). Managing project culture: The case of environ megaproject. *International Journal of Project Management* 25(3), 290–299.

van Marrewijk, A., and Smits, K. (2016). Cultural practices of governance in the panama canal expansion megaproject. *International Journal of Project Management* 34, 533–544.

van Marrewijk, A., Clegg, S.R., Pitsis, T.S., and Veenswijk, M. (2008). Managing public–private megaprojects: Paradoxes, complexity, and project design. *International Journal of Project Management* 26(6), 591–600.

Verweij, S. (2015). Achieving satisfaction when implementing PPP transportation infrastructure projects: A qualitative comparative analysis of the A15 highway DBFM project. *International Journal of Project Management* 33, 189–200.

Vicente-Oliva, S., Martínez-Sánchez, A., and Berges-Muro, L. (2015). Research and development project management best practices and absorptive capacity: Empirical evidence from Spanish firms. *International Journal of Project Management* 33, 1704–1716.

Voss, M. (2012). Impact of customer integration on project portfolio management and its success: Developing a conceptual framework. *International Journal of Project Management* 30, 567–581.

Williams, N.L., Ferdinand, N., and Pasian, B. (2015). Online stakeholder interactions in the early stage of a megaproject. *Project Management Journal* 46(6), 92–110.

Winch, G.M. (2013). Escalation in major projects: Lessons from the channel fixed link. *International Journal of Project Management* 31, 724–734.

Wood, V.F., and Bell, P.A. (2008). Predicting interpersonal conflict resolution styles from personality characteristics. *Personality and Individual Differences*, 45(1), 126–131.

www.ama.org/Pages/default.aspx,

www.ama.org/the-definition-of-marketing/

www.euromonitor.com/.

www.nielsen.com/fr/fr.html; for further examples, see also opinionresearch.com, google.com/intl/fr/analytics, stat.gouv.qc.ca, marketingpower.com, dialog.com, or lexisnexis.com.

www.nytimes.com/ 2002/03/24/world/under-pressure-chinese-newspaper-pulls-expose-on-a-charity.html,

www.pmi.org/,

www.standishgroup.com;

www.statista.com/statistics/682784/transport-infrastructure-global-projects/

www.worldbank.org/projects.

www.zenithmedia.com/ Groupe Publicis: www.publicisgroupe.com/fr

Yu-Chih Liu, J., and Rizki Yuliani, A. (2016). Differences between clients' and vendors' perceptions of IT outsourcing risks: Project partnering as the mitigation approach. *Project Management Journal* 47(1), 45–58.

Zhai, L., Xin, Y., Cheng, C. (2009). Understanding the value of project management from a stakeholder's perspective: Case study of megaproject management. *Project Management Journal* 40(1), 99–109.

Appendix 1: A Short Description of the Exercises Used during Our Seminars

1. A.F. Thériault: This Canadian boat manufacturer is producing unique boats for the American army.
2. African Warehouse: Contrasting views between stakeholders regarding the building of a warehouse in a poor neighborhood in Africa.
3. Algues Acadiennes: This Canadian enterprise sells unique sea products, in large part to Japan.
4. Amandine: This French artist is contemplating various ways to sell her art.
5. Art Gallery: This Canadian gallery is facing changes in its environment, forcing it to review its marketing strategy.
6. Boulangerie Benoist: This French family bakery is facing the arrival of a major competitor, right next to its store.
7. Casa Victoria: This French business provides short-term rentals, a concept between Airbnb and regular hotels. A new competitor is now challenging it.
8. Comeau Sea Foods: This Canadian business has ambitions for growth.
9. Fellini's Cat Litter: They are creators of an innovative product, but they are finding it difficult to market.
10. Gourmandises du Palais: The owner had different options to invigorate his store, one of which was to redesign the floor plan. The project crashed.
11. Haiti Hospital: After the earthquake, a Canadian company was hired to rebuild a hospital, but cultural differences soon put up obstacles.
12. Louis XIV: This Canadian IT business relies on growth through tailored customer relationships.
13. Magex: This Canadian IT business designs custom-made programs for the house/apartment rental market.

14. MEP Pharmacy: This US pharmacy wants to redesign its floor plan to improve the flow of customers.
15. Montréal International District: MID was a major undertaking to revamp part of downtown Montréal, Canada. Winner of 30 major national and international awards.
16. Ottawa *versus* Gatineau: Two nearby Canadian cities compete to attract tourists.
17. Québec Multifunctional Amphitheatre (Canada): A $400M CAD (Canadian dollars) project that was completed on time, within budget, and in full accordance with preset norms of quality.
18. Recycl'Art: This art festival promotes the use of recycled materials to produce works of art, and is open to the public every summer in Gatineau's parks (Canada).
19. Seacrest Fisheries: This Canadian plant changed focus and built a new plant to produce a new line of products.
20. Sentinelle: This Canadian healthcare provider is debating how to reorganize its two facilities.
21. Sherbrooke Toyota: This Canadian car dealership wants to keep ahead of the competition.

Appendix 2: The Core Psychodynamic Model

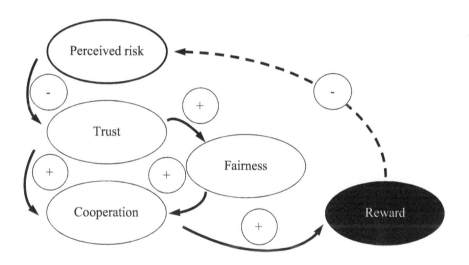

Appendix 3: Arguable Errors Found in Marketing and Project Management Literature

We herewith list the various "arguable errors" we uncovered in various books on marketing and project management. We believe this list serves the interest of science, and of practical application of theories, to challenge preconceived ideas. We do not pretend to hold the truth and may actually be wrong at times, but at least we foster discussion in our attempt to merge two disciplines: marketing and project management.

Chapter	Preconceived Idea	Our Challenge
1	Consumers' purchasing process	Not all consumers have yet "consumed"
1	Market agents include influencers and distributors	In fact, influencers are providers of opinions and distributors sell their services (and are thus suppliers)
1	Markets are a territory where companies compete	Monopolies prove otherwise
1	Omitting the step of purchasing in the purchasing process standard model	If there is no purchasing, then there is no purchasing process

(Continued)

Chapter	Preconceived Idea	Our Challenge
1	Products and services are one and the same.	They, in fact, display contrasting characteristics.
1	Project management is about satisfying customers.	Not at all; it is about meeting deadlines and respecting budgets and preset norms of quality.
2	Demographic segmentation.	It is more accurate to use socio-demographic segmentation.
2	Marketing consists in bringing value to customers.	Many products are put in the market that bring no value to customers.
2	Not considering perceptual and mnemonic effects as part of differentiation.	Perceptual and mnemonic effects assist greatly in differentiation.
2	One can segment products.	Segmentation concerns people, not products. Products are positioned and categorized, not segmented.
2	Perceived value and added value are one and the same.	Perceived value is in the eye of the customers; added value concerns the suppliers/manufacturers.
2	Perceived value increases according to what customers gain (benefits) and what they give.	Perceived value increases according to what customers gain (benefits) and what they give up.
2	Project feasibility is equal to project sustainability.	These are two different concepts. Feasibility only means project managers have the forces and means of actually materializing the envisioned product (deliverable). Sustainability (or, in a large sense, viability or profitability) refers to what happens after the managers have completed the project.

(Continued)

Chapter	Preconceived Idea	Our Challenge
2	Socio-demographic and geographic segmentations are one and the same.	They are two different pillars of segmentation.
2	The opposite of satisfaction is dissatisfaction.	These are two different concepts. One may not love someone, but that does not mean one hates that someone.
3	Feasibility of a project is equivalent to viability.	Viability means the project can last financially.
3	Ignoring points-of-no-return and points of autonomy.	Not a good idea.
3	Ignoring the fact that all projects must be measured.	All projects must be measured.
3	Ignoring the role of sources of energy and infrastructures.	Sources of energy and infrastructures must be considered.
3	Key failure factors (KFFs) and key success factors (KSFs) are mere opposites.	They are not the opposite to one another.
3	Project management books not discussing key consensus factors (KCFs).	KCFs are instrumental in the success of projects.
3	Psychological pricing makes it easier for customers to buy products.	Often, it is the opposite.
3	The iron triangle consists of timeline, budget (or costs), and scope.	Scope is not measurable. The iron triangle consists of calendar of tasks and activities, budget (costs), and norms of quality.

(Continued)

Chapter	Preconceived Idea	Our Challenge
3	The iron triangle is composed of delay, costs, and scope or norms of quality.	Delay is the last thing a project manager wants!
3	We can treat the human aspect of projects with little regard.	Quite the opposite. People are at the heart of points of vulnerability (POVs).
4	Ignoring measuring instruments.	They must be considered.
4	Managers can control risks.	Managers can only anticipate and prepare for risks.
4	Risks and vulnerabilities are one and the same.	They are vastly different.
4	Using SWOT (strengths, weaknesses, opportunity, and threats).	One should use SVOR (strengths, vulnerability, opportunity, and risks).

Index